The
Multidimensional
Traveler

The Multidimensional Traveler

Finding Togetherness, or How I Learned to Break the
Rules of Physics and Sojourn Across Dimensions and Time

By
Khartika Goe

New Page Books
A Division of The Career Press, Inc.
Pompton Plains, NJ

THE MULTIDIMENSIONAL TRAVELER
EDITED BY KIRSTEN DALLEY
TYPESET BY GINA SCHENCK
Cover design by Lucia Rossman/Digi Dog Design
Printed in the U.S.A.

To order this title, please call toll-free 1-800-CAREER-1 (NJ and Canada: 201-848-0310) to order using VISA or MasterCard, or for further information on books from Career Press.

The Career Press, Inc.
220 West Parkway, Unit 12
Pompton Plains, NJ 07444
www.careerpress.com
www.newpagebooks.com

Library of Congress Cataloging-in-Publication Data

CIP Data Available Upon Request.

To the two most amazing and remarkable individuals in my life, Akong and Ahma, my grandfather and grandmother, whose unconditional love has successfully enabled me to surmount any obstacles and hurdles along the way.
....

Acknowledgments

My utmost gratitude to my coauthor, Katy Kara; indeed, we wrote this book together and in Togetherness. I never thought that a friend of mine whose hobbies vary from getting manicures and pedicures to straightening her hair would become my dearest and most spiritual friend. Your friendship has been one of the biggest blessings in my life. Truly, no words could ever express my gratitude to you. We make the best team!

To the positive beings of the Togetherness, without whom this journey would be an endless tunnel. You have served as the many lights within my soul.

To my grandpa, for staying true to his feelings despite the logicalities, and for his unwavering faith and unconditional love. There are no words in this world that could possibly describe just how much you mean to me.

To my grandma, for her unconditional love, and for giving me the sweetest and funniest memories of my life. It was my good fortune that a string connected me to you.

To Laurie Kelly-Pye, Michael Pye, and Adam Schwartz—thank you for taking a leap of faith. Your assistance in the manifestation of this book, your positive energy, and your kindness have been felt from the beginning and throughout.

To Kirsten Dalley, my editor—thank you for your wonderful editing, continuous hard work, and incredible spirit.

To Ruth Parnell, for all our telepathic communications and your unwavering support.

To S.Teo, M. Mansur, and HG—thank you for your friendship, your support, and your kindness.

Many thanks and gratitude to my family, Duncan Roads, Dr. Harry Oldfield, and Evy King.

And most certainly to the entire New Page team, whose dedication and hard work have assisted in the manifestation of this book, along with that of many other brilliant offerings.

A Note to the Reader

All the dialogue in this book is presented accurately, word for word and unaltered, via an inexplicable process of assistance, which I will explain in a very general way in this book, and hopefully in greater detail in other books to come. Everything you read here occurred exactly as I describe it, without a single deviation.

—Khartika Goe

Contents

Prologue:
An Awakening

....

I want to learn to listen from the blind,
to sense from the mute,
to perceive from the deaf,
and to seek from the limbless.

Those of the crowd do not teach me the ways of the world or the
truth that lies behind the mysteries of life.
While those who are unique make my heart sing in joy and my mind
roam in endless wonder, for they are the true teachers.
They hold the keys to the endless doors that the crowd has tightly shut.
And it is when I surround myself with them that my true being
resonates melodiously.

And though I am aware that for many of them, this journey has
been both strikingly arduous and dispiriting,
it is a truth of the universe that a candle in a dark room can never
burn out on its own.

For a candle to burn out, there needs to be a wind,
and for the wind to come through,
a window must be opened,

....

and an open window will not only allow the wind to come through,
but also the light.
The light that further illuminates the dark.
You are the candle that illuminates this dark room.
A candle that will not only illuminate itself,
but all within the room.

Embrace the fact that you are different,
that you are more in tune with your energetics than your physical body.
Turn this physical vulnerability into a limitless energetic ability and be of
service to this universe.

Teach the ones of this third-dimensional physical realm of their true
energetic nature. Help raise the overall frequency of this world and ignite
the flame in others, as this is the only solution and all the help this world
needs presently—
an awakening to our true energetic origin.

Introduction

As a teenager, I would often visit astrologers, monks, Zen masters, and other individuals of the like whose various claims and perceptions would always overlap in the prediction that whatever I was to do in the future, I would be traveling quite a lot. Because of this, I had always assumed that whatever profession I ended up choosing would take me on various trips throughout different places of the world. And in fact, this turned out to be the case.

When I was in London visiting my high school friends, I traveled in a way that I had never dreamed possible. A true aficionado of exploring new cities and countries, I embarked on my first *multidimensional* travel. It was my third night in London and I had just returned back to my hotel room and retired to bed early, as I was still rather jet-lagged.

In that twilight stage somewhere between waking and sleeping, I suddenly felt myself being led or pulled out of my physical body by a strong, unseen force. I was abruptly transported to an unfamiliar, rather odd room. I was initially struck by its shape: It seemed to be composed of layers of geometric shapes, and it became apparent to me that this room had been constructed with the knowledge of sacred geometry. The room itself was almost entirely white, apart from a long, light-brown bench boasting a vista of a large glass window and whatever was beyond. As I gazed through this window from a distance, I could see a breathtaking garden. It had an incredibly natural, wild feeling to it, quite like that of a rainforest, yet all the plants seemed so meticulously and beautifully organized that I knew it had been planned and cultivated. I was in inexplicable awe of the wholesome beauty of this garden and each plant and flower in it; I had never seen any plants or flowers on planet earth that looked so alive, or that presented such unique colors, so vibrant and lively. My eyes remained fixed on their elegant movements; it was as though they were in a constant embrace and communication with one another.

Transfixed by their beauty, my attention was broken only when I heard a warm voice say to me, "Khartika, have a seat." The voice itself was not audible; it was as though I had heard it through my mind, not my ears. As I proceeded to walk toward the brown bench to sit, I noticed three very tall *light beings* standing next to each other. They looked so similar that I could not tell the difference among them. I noticed that they exuded a very peaceful and loving collective presence. Once I took a seat, one of the light beings joined me on the brown bench and said to me, "Do not be alarmed. We are a force of we and us. We do not speak individually, but in unison instead." Naturally my thoughts were racing as I tried in vain to comprehend where I was and what he had meant by "force of we and us."

Unknowingly to me, he read my thoughts and replied, "We are in a state of a much higher vibrational frequency. It is of a much higher dimension. The nature of we is a collective factor of unity, free from the bias of the 'I' or the ego."

I turned my head to gaze at him, thinking to myself, *He must be so pure. That is why he is so bright, even brighter than the sun. I cannot even look at him. Maybe I'd be able to look at him using sunglasses?*

Embarrassingly, he responded to my silly thoughts, "We laugh upon hearing your thoughts. You must remember that the humor in life is important, the laughter. Our existence is of a pure force, and we are based in a pure dimension of only wisdom, love, light, and knowledge."

Deeply curious, I proceeded to question him: "Pure force? Does this mean that you cannot exist or come to earth?"

He answered, "We are unable to manifest there, as the pace of the vibrational frequency does not match ours." While listening to him, I couldn't help but notice a white flower in the garden that resembled a hybrid between a lily and a Venus fly trap. It began to move back and forth in an odd manner and then dripped a bright, golden-colored dew.

The light being continued to respond to my thoughts, answering, "The gold that you see is its ether. The color is gold because the essence has reached a high spiritual evolution." Noting my confusion, he continued, "The flower that you see is the soul of its existence, while its ether is its energy."

A smile flitted across my face and I asked, "So, sir, you are not made up of molecules as we humans are? Are you made up of light particles instead?"

I sensed that he smiled, too, as he replied, "Everything within this universe has an ether. Ether is the primary and most basic component of the universe. It is an energetic force derived from the universal force collision of intergalactic realms. It is an energy of constant and absolute state which does not fluctuate or dissipate through time. Ether existing within the human mind itself is a tunnel of force interconnected with the heartbeat."

As he finished this sentence, the other two light beings moved and stood behind him. I looked at them curiously and asked, "Sir—I mean, the you guys force—I mean, the we force, sir... Oh, I don't know how to address all of you. I guess, the we force sir. I don't understand—what exactly is this ether that you are talking about? I mean, that *you all* are talking about?"

The light being seated next to me was silent for a moment before he replied, "Ether is energy. When one is speaking of the etheric realm, one is speaking of a differing layer of existence that exists in the human realm, not the physical in which many reside. The etheric realm is an existence of energetic composure. Unlike humans, whose lives simply revolve entirely around the physical realm and who are satisfied with just the physical stimuli of sensations and perceptions, many other beings of the intergalactic realms, in addition to using their souls, have turned to their energetic bodies as a means of communicating with each other. Governing the energetic body happens only after mastering the mind— the mind being the steering wheel. This is an automated action-reaction. Being a master of the energetic is being fully conscious of the energetic body and utilizing it to its full potential."

Suddenly, he turned to me, his gaze locking onto mine, as though anticipating my next question: "Sir, does anyone on planet earth know of all this?"

He quickly responded, "Many adepts of your world in the past have successfully utilized this knowledge and used their energetic bodies as a means of communicating with us and other beings. They have also successfully traveled to the vastness of space. Khartika, every existence on every realm exists with the energetic dimensional layer. It serves as the true hologram of any particular realm itself; we are referring to earth specifically. You must trust us when we say that beings of other realms are fully conscious of their multidimensional existences."

To this I replied, "So, sir, are you an energy too? Do you have the ether, too?"

"We have completed the unity of our energetic body with the soul. But the two have always been united for us. Such is the nature of the we. Unity." I was visibly in utter confusion and at a loss for words, so he kindly added, "You have to take everything step by step, not all at once. It will be too overwhelming. You will not remember all of our conversation, only some. What you remember will be enough to share with your kind. Some will be alarmed by what you have to share, but you must just allow it. Never fear, for fear is the worst destruction of the self. We are always within you, never fear."

Confusion further clouded my thoughts, and I felt undeniably and irretrievably lost. Desperately craving understanding, I asked him, "I don't understand, sir. Who are you, really?"

He replied, "You *won't* understand, for now it is not the time for you to fully understand. Many exciting experiences await you on your journey. Asking us who we are is similar to being in a forest with many trees and asking us which tree we are. Do you understand?"

I simply nodded in response, accepting this answer, though more than anything I wanted to know more.

After this we had further conversations, which I cannot recall. The last thing I remember was him asking me, "Do you promise that you will do this?" to which I replied, "Yes, I promise." And upon stating my promise to the light beings, I felt myself being propelled by a whirlwind-like force against my chest that sent me back into my physical body within an instant.

Upon returning to my physical body, I opened my eyes and quickly grabbed my watch. It was 5:03 a.m., which meant I'd been asleep for seven hours or so. I flung open my laptop and, determined to find any answers to the countless questions galloping through my mind, I proceeded to Google keywords related to my experience. To my dismay, the results were overwhelmingly disappointing and frustrating, most of them suggesting that a drug-induced hallucination had been the cause of my experience. I have never touched drugs, nor do I intend to ever use them. I was not drinking at the time, either, as I just don't drink. Hence, the previously always-dependable Google disappointed me that early morning and left me alone to ponder on what had happened to me as the sun came up.

Before this enlightening experience, my interests had solely involved cooking and traveling around the world, not multidimensional realities! This experience jolted my belief system with such incredible force that I knew I would never view the world in the same way again.

Though you may feel accomplished and content living a life revolving entirely around material achievements, I can assure you that you will never feel *complete* until you learn to seek the truth about our universe. After taking a first bite from the apple of knowledge, your world, no matter how small it seemed before, will never be the same again. It is an exciting yet challenging journey. The decision is yours: whether to go forward or live in oblivion. Many have claimed that ignorance is bliss, but what good is bliss if it's merely temporary? One day you, too, shall die, and what else will you be able to bring to the other side besides knowledge?

Understanding that the universe is indeed subtle and that within you is something that has been written about in countless sacred texts, requires a spark of curiosity within your own self. Stop the timing of the clock, for it is but a man-made illusion of hastiness, and instead focus on the virtue of patience. Patience and other good virtues are important stepping-stones on the path to multidimensional self-discovery. After all, your very nature is multidimensional, and you are living in a multidimensional world—whether you realize it or not!

In order to ascend a ladder, you must carefully align your feet on each step before proceeding to the next. By opening this book, you have taken the first step and have reached a point where you must make the decision of either continuing up the ladder, or retreating back to the ground. The choice of whether or not you will discover your energetic abilities is entirely yours to make, for fate is but a door flung open by an unseen wind, while a journey is a self-pronounced opportunity that surfaces to us upon feeling fate's touch. As everything within our universe is dictated not only through fate, but also through free will, you must make this discovery on your own. Your decision to do so will only receive the universe's blessing upon your soul's approval.

1

The Fundamentals of Multidimensional Travel

What if I told you that the concepts of both time and space were created by your own dogmatic mind? What if I told you that by altering your perceptions of both time and space, you would be able to disregard their existence entirely and travel freely through time? The concept of multidimensional realms is one that is difficult to accept, and even more difficult to understand. A bundle of strings can serve as a useful metaphor here. If you looked at a bundle of strings of different colors and thicknesses from a distance within a dimly lit room, you could easily mistake the bundle for a single rope. Only after careful examination in a well-lit room would you realize that what you had identified earlier as a rope is in fact a group of strings that, when placed together, appear as a single bundle. This same idea applies to the concept of multiple dimensional existences. As citizens of the physical realm, we are only exposed to our physical surroundings and are entirely dependent on the perceptions of our senses; as a result, we miss the many different realities that exist on our planet. Each and every one of us possesses the natural ability and knowledge needed to communicate with realms that exist beyond the physical dimension. However, the odds of our doing so successfully depend entirely on how we choose to use the power of our mind!

Traveling multidimensionally can be achieved through full or partial awareness. Multidimensional journeys taken in full awareness are not a new thing. In fact, being able to travel multidimensionally was regarded as quite normal by many ancient civilizations. Possessing knowledge far beyond our own, many ancient civilizations were actually able to predict their fall and eventual disappearance, and conveyed this spiritual knowledge through their art. Understanding the potential dangers of conveying this kind of knowledge to those who lacked spiritual understanding, these civilizations imbued their art with various spiritual codes that could only be unlocked through the elevation of the viewer's spiritual consciousness. It is only when someone's spiritual vibration is similar to that of the work of art itself—whether a painting, sculpture, or architectural creation—that he or she will be able to discern its hidden knowledge.

You may well ask: Why use art and not words to communicate? We all know that language differs by region and changes constantly through time. These wise ancient civilizations were well aware that the only way they could convey their knowledge universally and without change or bias across time and regions, would be through works of art. Furthermore, the use of language requires the engagement of the left brain, which is achieved and elevated, in part, through education. On the other hand, interpreting art requires the use of the right brain primarily, an activity that is quite universal among all human beings despite their varying levels of education. The fact that these civilizations chose to convey spiritual knowledge through art serves as further proof of their advanced wisdom.

These works of art, when looked at with an open mind, often contain depictions of and references to otherworldly beings. These beings often have angelic or, conversely, demonic characteristics and appearances. Most multidimensional travelers know that these two very stylized depictions are indeed portrayals of the two types of beings that exist in the higher and lower vibrational realms, respectively. During your first multidimensional journey you will probably encounter beings that inhabit the lower vibrational existences. As these works of art so graphically depict, these beings occupy the negative vibrational range; they will resemble anything from a sorrowful and broken human being to an extremely grotesque creature.

Within the lower-dimensional realm, there exist three different levels. The lowest of these contains those soulless entities whose consciousness has been completely lost through time. The middle level is inhabited by those earthbound souls who are ensconced in clouds of their own negativity and hence have lost a part of their consciousness. The highest level is that of the *earthbounds*. These beings are most similar to what we think of as ghosts. Earthbounds have lost their physical bodies and instead roam around the human energetic sphere in their energetic bodies. Most of them have, in fact, experienced sudden or traumatic deaths and thus find themselves lost in time and simply existing in the limbo of space.

Unfortunately, most multidimensional travels take us to these lower-dimensional realms. The reason for this is that the most common human vibrational frequency resonates with and even matches that of the beings in these realms. Yes—sadly, our current society greatly lacks spiritual understanding. Most people just naturally travel to these three different lower levels; in fact, 99 percent of all people have probably visited at least one of these three realms in their sleep, most of them subconsciously and with little to no recollection at all.

Earthbounds, whose "bodies" are composed of energetic molecules, typically roam around the human sphere completely invisible. But sometimes they can be seen. People see ghosts most often when they are in the state of exhaustion, due to the alteration of the vibrational frequencies of their energetic body. This phenomenon is an important quality of our multidimensional nature. Often, when our physical body is not entirely energized or even in a state of alarm or danger, the energetic body will automatically escape the physical body as a means of survival. A great metaphor for this is driving a car. If the car runs out of gas or is involved in an accident, the driver usually gets out of the car (or is forcibly removed from it). As you can see, the ability to travel out-of-body is a natural, even biological, phenomenon.

Electrical Encounters on the Island of Bali

Energies exist all around us; in fact, we are all made up of energies. In a living individual, the human energetic body is connected to the human mind. Though the energetic body is of a subtle form, it is composed of vibrational frequencies. In the cases of earthbounds who have detached themselves completely from their physical body, they are left to roam around freely in their energetic body. Thus, when they want to gain the attention of a living person (or simply cause a nuisance), many will turn to the manipulation of electricity. The following is an example of this phenomenon when it happened to me.

I was on a quiet vacation on the beautiful island of Bali, Indonesia, to celebrate my 19th birthday. Two friends and I were sharing a room for the first night. I was lounging on the couch and thinking about going to bed when, at exactly 12 a.m., I heard my friend Lily scream in the bathroom. I rushed into the bathroom only to find her exclaiming repetitively that her hair dryer had turned on automatically by itself. Our other friend, Mary, proceeded to laugh it off in a sarcastic tone, "Yeah, right!" As I was examining her hair dryer, the television in our living room suddenly turned on by itself. When this happened, my friends clung to one other, absolutely terrified. I, on the other hand, ran to the living room to witness this paranormal experience up close.

I stood right in front of the television as the channels proceeded to switch every second and as the volume jumped from low to high and back again. I smiled calmly and communicated telepathically through my mind, *I do not fear you. The only thing to fear is fear itself. Please stop this.* I had experienced such incidents countless times before; this time I was prepared. Indeed, the earthbound spirit heard my request and stopped the manipulation of the television set for the next 20 minutes.

I took this opportunity to calm my friends and lead them out of the bathroom where they still remained in fearful hiding. I lied to them saying that it was probably just a result of malfunctioning circuits, but Mary, a Catholic, insisted that we had been visited by a ghost. A frequent hotel hopper, she proceeded to open all the drawers in search of the hotel's complimentary Bible, but much to her dismay was only able to find a copy of the Hindu *Mahabharata* in a drawer next to the bed. To my surprise, after flipping through the pages, Mary seemed rather relieved and exclaimed to us, "Same thing! Same damn thing! All religions are the same!" and began to read phrases out loud from the "Bhagavad Gita" section, in which Krishna tells Arjuna of the nature of his being. I looked at her in astonishment, my mouth and eyes wide open, struggling to fight my laughter.

Unfortunately, this peaceful state did not last long, and the earthbound soon went back to making its mischief, this time attacking all the lights in the room. The lights started to flicker on and off, as if in response to the entertainment the entity was receiving from my terrified yet comical friends. Mary's relief suddenly vanished, and in the midst of the flickering she dropped the copy of the *Mahabharata* with a loud thump and darted out of the room. My friend Lily quickly followed her, both forgetting that they were dressed only in their pajamas. After they had left, I shook my head in disappointment and desperately exclaimed out loud, "Really! Please! Stop it. Listen, I'll help you, all right? I can't see you, unfortunately; I'm not a medium in that sense. I promise tonight I'll help you, but just stop this!" At this, the mischief seemed to cease for a few minutes—at least until the phone starting ringing over and over. I simply sat there ignoring it, thinking it was probably the earthbound stubbornly manipulating the phone.

After about 15 minutes of this I heard knocking on my door. My curiosity getting the best of me, I quickly stood up to get the door, only to find a concerned hotel manager standing there sheepishly. He was about 5 feet tall and was dressed in traditional Balinese attire. He apologetically informed me that my friends had been moved to another room and that he would lead me there, as well, but I assured him that I was not the least bit shaken and that I would actually *prefer* to stay in my room. A puzzled look came over his face and he continued to apologize for the inconvenience my friends and I had experienced, and then began telling me of the possible short-circuit that caused this. I was still thinking of the earthbound when suddenly he dispensed with the formalities and confided, "We Balinese believe in spirits roaming around. We live in harmony with spirits. So we give them offerings, but sometimes..." he trailed off, "There are evil spirits roaming around. Tomorrow we will call someone to cleanse this place. You sure you want to stay here?" I replied reassuringly, "Yes, don't worry. I know what you are talking about. I'm familiar with this kind of stuff!"

I said goodnight, gently shutting the door behind me, and got ready for bed. But before I allowed myself to drift off, I knew there was a message I had to get through to the earthbound. As I turned off all the lights, I spoke clearly into seeming emptiness, "Don't worry! I remember my promise. As I am in the state of falling asleep, talk to me, okay? Because that's the only way that I can hear you!" And sure enough, as I drifted off, I suddenly heard a little girl's murmuring voice echo through my mind, "Can you hear me? Can you hear me? Hello?"

I replied to her telepathically, "Yes, my name is Khartika. What's yours?"

I could clearly make out her elated relief as she replied with enthusiasm, "My name is Stefanie. I'm 11 years old."

Delighted to have finally met my midnight visitor, I curiously asked her, "Stefanie, are you from Bali? What happened to you?"

She started to reply, "Yes, I'm from Bali..." but before she could finish what she was saying, she was cut off by an old man's raspy voice.

"Hello? Can you hear me, too?" he croaked, desperation evident in his voice.

I telepathically told him that I could, and I heard Stefanie's warm voice resonate in my mind again: "This is Pak Tejo. He's my friend." (*Pak* means "sir" in Indonesian.) After hearing her say this, I felt my body drift into a state of utter exhaustion, intensified by a sudden onset of body aches. As an avid multidimensional traveler, I was well aware that I was about to detach from my physical body and be transported into a lower-dimensional existence. I hastily warned my new acquaintances of this in advance: "Pak Tejo, Stefanie—give me one second and I'll be able to see you." Within an instant, I found myself face to face with them.

Stefanie was a beautiful young girl with long, jet-black hair and tan skin. Pak Tejo, on the other hand, was completely terrifying. He looked to be at least a century old, and he had no arms, no hair, and an empty left eye socket. He grinned at me excitedly, exhibiting a disquieting row of fake gold teeth. When he noticed me staring at his eyeless left socket, he explained matter-of-factly, "This eye was popped out by the bastard who killed me. I curse him to this day! I will make him pay. He took it out and placed it on an altar before killing me."

I was both shocked and curious. I asked him hesitantly in a quiet voice, "On an... altar? Why, Pak Tejo?"

His reply was filled with hateful anguish: "In Indonesia many people want to acquire powers through black magic rituals. The belief has it that if they kill 100 people and present their eyes on an alter before their evil god, they will acquire the powers." His accent was thick and his reply immediately reminded me of the movie *Indiana Jones and the Temple of Doom*.

Without thinking, I ignorantly asked him, "Such... things. They still exist? But how do they work? It doesn't make any sense."

He was clearly offended by my thoughtless comment, replying in a noticeably raised voice, "Yes, of course it does. You think I'm lying? You're here right now and you're alive in your physical body, aren't you? If you tell most people, they'll say, 'Is it really even possible to get out of your physical body?' Same thing. I will tell you what I know. If you don't believe me then ask the villagers in Indonesia, especially in Bali—not the modernized city people from Jakarta who reside in Bali, but real villagers from Bali or Java! Based on what I have learned, we all get stuck in this lower state after getting murdered in brutal ways. They use chants, and their gods are darkness!"

I was so focused on Pak Tejo that I had almost forgotten about Stefanie. She asked in a soft tone, "But Pak Tejo, I thought you are not from Bali, but from Karang Asem. And that you lost your arms from an accident."

Pak Tejo turned to Stefanie with his golden, toothy grin and replied, "Little girl, that happened to my arms and I *am* from Karang Asem! But I didn't tell you the truth of how I died and my left eye, did I? Because I did not want to scare you!"

Stefanie retorted defensively, "But I am not scared at all. I've heard worse! I told you the truth of what happened to me, Pak Tejo. You should've told me the truth! Mine is scarier than yours anyway."

Desperately wanting to know her story, I asked her what had happened to her. She looked at me rather sadly, a sorrowful recollection evident in her eyes, and, after a brief pause, proceeded to tell me her story. "My parents have many children to help them with their rice fields. I am the youngest and when I was alive, I had mind and walking problems. I couldn't walk, talk, or think properly like I do now. So, they were embarrassed of me. And in our village having a child like me is considered a bad omen, like a curse. So they kept me locked in a small room. My horrible brothers used to torment me, too, and hit me. I hate them. But my death was caused the year famine reached our village. They decided not to feed me, and one day I found myself floating above my body and heard my parents lying to the head chief of our village, telling him that I had died because of a disease. Ever since then, I've been coming to them so that they'll feel guilty for what they did to me."

Her story deeply saddened me and I said to her softly, "I am sorry to hear that, Stefanie. They will eventually learn from what they did." I paused briefly and continued, "But don't you ever wish to go to a better place?"

"Pak Tejo told me we can't and we are stuck here forever," she explained.

I confronted Pak Tejo sarcastically, "And why is that, exactly, Pak Tejo?"

He replied to me as though the answer were obvious: "Because we missed the light and our type of death is not normal. If it is true that we can go to a better place, why haven't we been able to? We've been stuck here! You don't understand," he said, shaking his head as he turned away from me.

"Pak Tejo, maybe I *don't* understand. But my life has not been what you deem as 'normal.' Me standing in front of you and conversing with you presently while my physical body is sleeping on the bed—is that normal, Pak Tejo? Please just listen to me, I've met people who've experienced quite similar things as the both of you—though I admit, not as horrible. But every life is simply a lesson and a cycle. You have to see it as a mere lesson, quite like a classroom. The both of you have had many lives before this one, too. Have you ever thought that maybe you experienced what you did in order to learn to forgive and move on? Or perhaps to understand this present state of dark existence and learn to break free of it? In my experience, breaking free of it requires a high level of emotional and mental control. There's a much higher existence out there that awaits you; I promise you both this." I noticed Stefanie's eyes immediately light up, and with a beaming smile and a delighted tone she asked me if I was sure about this. I nodded in reassurance, "Yes, Stefanie, and you will be able to reunite with people you love that have passed away."

Her smile grew and she practically yelled excitedly, "I'll be able to see my great-grandmother?" I told her it was all entirely true, every word I had spoken.

Pak Tejo suddenly interrupted us with his booming voice, "I will not go anywhere until that bastard who killed me pays for what he did! I curse his existence!"

I sighed and explained to him, "Pak Tejo, from the experiences I've witnessed, you will not be able to elevate to a higher-dimensional existence if you do not let go of all and everything. I told you: this life is just one of the many you've had. Everything is just a learning experience."

He replied furiously, "What do *you* know? You have no idea what we have been through. You are just a walk-in!"

Stefanie said calmly to him, "Pak Tejo, I want to see my *nenek* [great-grandmother]. I'm tired of being here."

Upon hearing her words, Pak Tejo looked down with his one eye. He was visibly embarrassed about losing his temper and muttered to Stefanie in a lowered voice, "Okay, go then, little girl. But do not forget to think of me, and you know where to find me if this know-it-all girl fails." Stefanie's smile reemerged, and Pak Tejo lifted his gaze to look at me with his one eye. "Good luck," he said to me archly, and turned his back.

I stopped him. "Pak Tejo, I have one request! Please don't go around scaring the living daylights out of the living with electrical mishaps in hotel rooms. And one more thing: You know you can find me if you change your mind, Pak!"

He simply clicked his tongue and mumbled, "Who the hell does she think she is?" and walked away into darkness.

I was now entirely focused on Stefanie and explained to her, "It doesn't happen like magic. Nor can I perform magic. You must be the one to do it yourself. I'll walk you through it." I paused and waited for her to nod as a sign of approval to proceed. As she nodded with a big smile, I continued, "First I need to ask you: Do you love your great-grandmother?"

"Yes, very much," she exclaimed. "She was the only one who was nice to me."

I instructed her, "Okay, think of the love you have for her. Feel it. Just focus on it entirely. Forget about your evil family, forget them, forgive them—it's all in the past. They were just a part of your learning experience. The state of the emotion and mind is everything, Stefanie. We are raising your vibrational frequency as we do this, and when your frequency doesn't match that of this existence anymore, there will be a door to a higher existence. Trust me! You must picture a much better place, a much higher existence where everything is pleasant and nice. Close your eyes if it helps." Following my instructions, she closed her eyes and I proceeded, "Everything here, as it is throughout the entire universe, is governed through the universal law. No help will be given to you unless you ask for it. So you must ask in your mind for help to take you to a higher place. Keep feeling love, and only love."

I had no doubt that she was listening to me and doing all that I was telling her to, because within an instant, I saw a small sphere of bright white light begin to approach us. As it traveled closer to us, growing in size, I instructed Stefanie to walk through it, daringly thinking to myself that perhaps if I went through it, I, too, would end up in a higher-dimensional existence. I excitedly followed behind her, thrilled at the prospect of another multidimensional adventure. But as I approached the light, I felt what seemed to be an invisible barrier preventing me from entering it.

I watched Stefanie successfully pass through the barrier, and the white light began to lift her higher and further away into the distance as her image contracted and diminished. I remained blocked by the invisible barrier, but through my squinting eyes I saw what appeared to be a young woman come out of the light and take Stefanie by her hand. Stefanie followed her, pausing only once to turn around and wave goodbye to me. Within a heartbeat, I felt a powerful, whirlwind-like force push against me and found myself pulled back into my physical body.

It was precisely 7:05 a.m. when I woke up. I had been asleep for approximately two hours. I stayed in bed for a while, grinning at the thought of the extraordinary journey I had just been on. Then, the phone rang. I heard Mary's concerned, hectic voice on the other side of the line, "Hello, did you just wake up? I couldn't sleep all night. And I don't want to go to your room, since I'm sure there's a ghost in there. Why did you stay there? You're crazy, you know that? Or is it that you don't believe me and think of me as crazy for thinking that it was a ghost? Or is it because I snore? Anyway, we're going to breakfast. I've already talked to the manager. We're changing hotels today. We're getting a one-bedroom villa with a cheaper price; you can sleep in the living room if you're scared of my snoring." Upon concluding her monologue she hung up. I proceeded to shower and prepare myself for breakfast.

Our breakfast that morning was incredibly awkward. I sat opposite my two friends, and after we exchanged our good mornings, they went back to their silent mode. I surreptitiously glanced at the two of them—eye bags, messy hair, and pale skin. They looked as though they had indeed seen a ghost! Thankfully, Mary soon broke the awkward silence. "Listen, I've been thinking about this. The reason that happened to us last night is because we have been doing truly horrible deeds."

I looked at her in utter confusion. "Horrible deeds? What horrible deeds did I do?" I had no idea what she was talking about.

Mary replied as though it were obvious: "We stole shampoos and soap bars from the housekeeping carts yesterday evening after dinner. And have you forgotten?" she asked with an ashamed look on her face. "We were planning to steal the hotel bathrobes, too! Stealing is wrong!" I could tell that instead of sleeping, Mary had been thinking about these "horrible deeds" the entire night, and unfortunately I found it hard to contain my laughter. She threw back at me in a half-confrontational, half-joking tone, "Ha, ha, ha—you think this is all *so* funny, that I am crazy for thinking it was a ghost!" She was convinced! I simply kept quiet and let Mary continue to hold forth on the relationship between our sins and the strange goings-on in our room. After she was done, I decided to go off on my own for a bit and enjoy a quiet stroll alone on the beach, which I enjoyed immensely.

After the three of us met up later and were in the car to go to our new hotel, Mary announced, "I met with a *ketut*—you know, a Balinese medicine man—and according to him, we are to leave our suitcases and everything we had in that haunted room out in the sun for 11 hours. It's to ward off the evil energy that is attached to it."

Lily exclaimed sarcastically, "So we should take off all the clothes we are wearing now and go to our hotel nude, since they also have an evil energy?" making air quotes around "evil energy."

Mary lifted her chin and answered plainly, "No. You either do or you don't; it's up to you. *I'm* leaving my suitcase out in the sun for exactly 11 hours as the *ketut* instructed." When she saw me grin in response, she barked, "You are *so* narrow-minded! You really don't believe that it was a ghost, do you?"

I continued to smile and returned calmly, "Ghost? What ghost? I stayed in that room all night. It was a malfunctioning circuit, obviously!"

To this day, Mary must think I am deathly afraid of her snoring.

Grandmother

There is a demonstrated and proven relationship between old age and disillusionment. A similar strong correlation has been found between young children and the ability to see imaginary friends. People who don't fall into either of these age categories tend to casually brush off these correlations and attribute them to the degeneration of the mind (in the elderly) and the power of imagination (in children). However, what they fail to recognize is that children are newborn souls, while the elderly are souls that will soon leave this physical world. Therefore, it makes perfect sense that both children and the elderly would be naturally inclined to be more energetically sensitive to the multidimensional worlds.

After my trip to Bali, I spent the rest of my winter holiday that year at my grandparents' home. I had spoken to my father prior to my arrival and he had updated me on my grandmother's health. Her Alzheimer's had worsened and she had been hallucinating. During the first two days I spent entirely with my grandmother, I recall her frequently mentioning the presence of a strange man or woman in the house. Even on my first night home, when we were alone, I remember her inviting a woman to sit with us for dinner. My parents shook their heads and exchanged looks of sympathy and mutual commiseration over my grandmother's mental state. I also wrote it off as a hallucination and continued to think so until my third day home.

I was sitting down next to my grandmother giving her a shoulder massage when she exclaimed loudly, while gesturing with her hand to nobody visible, "Don't just stare at us and stand there like a statue! Just sit on one of the couches!" Almost immediately, I saw her nurse shaking anxiously as she beat a hasty retreat to the kitchen. Curious, I asked my grandmother in a soft tone, "Who is it, Grandma?"

My grandmother snapped back, "An old man; he's annoying me. Just standing there like a statue and smiling."

"What does he look like, Grandma?" I questioned her lightly, pretending to be only casually curious.

She answered with a sour look on her face, "Tsk, tsk... nasty! Like he just got into an accident! No hands, one eye, not a strand of hair, and very skinny, like a dried anchovy, as though he only eats cucumbers for breakfast, lunch, and dinner. He gives me the chills!"

I stared at her in shock and quickly asked, "Does he have gold teeth?"

"Yes! He's annoying me! All he does is stare at us!"

I gulped in disbelief. I knew that Pak Tejo had come to speak to me and, based on my previous experience with him, I was afraid that he would cause a nuisance in my grandparents' home. Fortunately, this time he didn't manipulate any of the electronics, but knowing his nature, I was well aware that he was very impatiently waiting to converse with me, so I prepared to travel early that night. As I was drifting into sleep, I addressed him telepathically, "Yes? Pak Tejo?"

I soon heard his now-familiar laughter as he replied in his gravelly voice, "Khartika! I have great news! You need to help me!"

I paused for a few seconds. I was dealing with Pak Tejo here, after all. "If it's not a negative thing, then yes, I will. What is it?"

Full of excitement, Pak Tejo replied, "My murderer is currently staying put in a house, and to my knowledge he will be there for the next two months. I will tell you the address and you can give it to the police. Then, I will follow your advice and leave, like Stefanie."

I let out a deep sigh of disappointment and replied, "Pak, it doesn't work that way. An eye for an eye does not work in your world. If I do that and you derive satisfaction from it, it does not help you learn your lesson or solve your problem. Your task is to simply let go and understand the illusion of the previous life you lived, that it was all part of the larger lesson at hand."

My reply clearly infuriated him, and he barked back at me, "I told you I will leave to a higher place after you *do* this for me!"

I tried again to get my point across to Pak Tejo, explaining as calmly as I could: "In that case, *Pak*, you won't be able to! Tell it to someone else, maybe a psychic! You will not go to a higher place if I do this for you. You can only go to a higher place if you learn to let go, so come back to me when you've learned to let go and I'll tell you what you can do."

This response didn't satisfy him either, and he shouted at me, "I came all this way! You are useless!"

I felt incredibly sad for Pak Tejo that, due to the unfortunate hardships he had experienced during his lifetime, his sole focus after death had consisted of revenge and hatred. I wondered whether I would have acted in the same way, had my life had been so violently cut short. I thought about it numerous times, and whether I had done the right thing in Pak Tejo's case or whether I should have written down the address he had given me and reported it to the police. But I had an irrefutably strong feeling that I was not to meddle in this case. I was well aware that if I had done so, Pak Tejo would not have learned the lesson from his previous life—the lesson of simply letting go.

A Lost Angel

After my month-long winter break, I arrived back at my apartment in Los Angeles. Normally I cleanse my home when I've been away for a week or more, as I don't want earthbounds constantly flocking my home for assistance. I have a life of my own in the physical world, after all, and I needed to remain focused on college and maintaining my grades. However, this time I was extremely jet-lagged and exhausted when I got back from the airport, and so I foolishly headed straight to bed without performing my customary cleansing.

As I was drifting into sleep (but still in full consciousness), I felt a light slap across my right cheek. I knew instantly that someone was trying to get my attention, so I quickly separated from my physical body. Almost immediately, I made out the image of a dark shadow of a woman running away from me toward my bathroom. I called out in desperation after the shadowy image, "No, don't run! I'm not mad at you. I just want to talk!" The shadow stopped in its tracks. I continued, "My name is Khartika. I'm curious, as I've never witnessed anyone appear to me as a shadow before. Why are you a shadow?"

"Because I exist in a lower dimension than you." A high-pitched voice of a woman arose from the shadow. I could tell she was nervous. She spoke quickly and succinctly, as though she expected our conversation to end at any moment. She must have noticed I was not very satisfied with her vague response, as she soon added, "By the way, you can call me Mimi."

I was still puzzled by what she had said earlier, so I responded, "But I've seen earthbounds existing in a lower dimension before and they never appeared as shadows to me."

Seemingly cautious, she stayed a few feet away from me and hesitated slightly before answering, "You must've been vibrating in the same frequency with them, or—" She paused again to think. "Or altering your frequency to suit theirs. Your frequency and mine don't match, so I appear as a lower form, a shadow, to you."

Struggling to wrap my mind around this, I asked her, "Well, are you an earthbound or not? What happened to you, if you don't mind my asking?"

She finally slowly approached me, as if she had realized that I would not harm her, and began telling me her story. "Well, yes, I am an earthbound. I died when I was 9. I was hit by a car in 1978 in Los Angeles."

I stood there puzzled—she certainly did not have the form of a 9-year-old girl, but rather that of a grown woman. I replied politely,

"Forgive my curiosity, but I have met other girls who died young, too, and they retained their youthful looks, but you don't seem to be 9 years old. You seem much older. How can that be? I thought earthbounds tend to look the same in life after death as they did in life."

"Even when it comes to earthbounds," she intoned, "there exist many sectors of dimensional existences. In my existence, it is the highest of the earthbound existences; our existence greatly mimics that on earth."

I was entirely lost, so I asked, "Are you trying to tell me that everything that exists in the physical life on earth, exists in your dimension as well?" Suddenly understanding, I went on quickly, "So this means that the concept of time, as it exists on earth, exists in your realm because you deem it to exist, and that maybe this is the reason why you age as one would physically? Because your mind is altered by time?"

"I guess," she said hesitatingly. "I mean, most people in my existence died suddenly, a few from self-mutilation."

Puzzled, I asked, "But why would you want to be in a lower existence or remain an earthbound, when you could go to a higher place?" As I concluded my sentence, however, I heard the discordant ringing of my alarm clock, and almost immediately I was pulled back to my physical body.

When I woke up, I sat up and said out loud (knowing that my new acquaintance would hear me), "I'm sorry for the interruption, but I have to go to school now. I have two art history lectures today, but tonight I promise that we will converse again. I'm not a medium, unfortunately, so if you try to talk to me while I'm in my physical body, I won't be able to hear you or talk to you, although I may be able to sense you. Please don't think that I'm ignoring you!"

That day, as I sat through my Buddhist and medieval art history classes, I experienced shock and disbelief as I discovered how nearly every work of art from the past depicts multidimensional realms and beings, yet most people still think of these things as purely imaginary or symbolic. I remember my Buddhist art history professor projecting a slide of a Tibetan work of art to the class and commenting on the different versions of hell the piece was depicting. I chuckled inwardly after hearing this, as I had visited some of those so-called hells in my travels! As I drifted into deep thought about Mimi, the earthbound I had met the previous night, and contemplated which dimensional existence she actually belonged to, my thoughts were distracted when I heard a girl sitting in front of me whisper to her friends about what creative minds these artists must have had to be able to "imagine" such scenarios. At the time, her comments made me feel both frustrated and disappointed. It became clear to me that although we live in a vast universe composed of many different galaxies, some people are too self-centered or narrow-minded to believe that anything exists beyond the physical world.

Later that day, I continued my normal routine of completing my schoolwork and cooking myself dinner. While I was working on an assignment, I again sensed the earthbound's presence around me. It was highly distracting; I even heard high-pitched sounds in my left ear. The earthbound's presence was strong and carried with it a sense of urgency, so I decided to put my schoolwork aside and prepare to travel. As I was falling asleep, instead of detaching entirely from my physical body as usual, I heard the high-pitched noises again and saw various images flash in front of me, as thought I were watching a movie. The images depicted a little blonde-haired girl who was obviously raised in a wealthy family, along with a twin sister with whom she was clearly very close. As I watched her story unfold, I learned that she had spent most of her days with her nanny and playing the violin. One day, while she was walking down the street with her twin sister and nanny, she and her sister were hit by a car, missing the nanny by only an inch or two. I saw images of the little girl floating above her body, staring down at her physical self. Then, I saw her twin sister in a wheelchair, obviously handicapped due to the accident.

Suddenly, I heard a voice echo through my mind: "Are you okay?" The images instantly disappeared, and I found the dark, shadowy woman

I had met the previous night standing in the corner. She was thin, with long, blonde, curly hair and blue eyes, and was dressed in a long white dress. I recognized that I was in my room—the layout was the same—but there was something slightly different about it. I was confused and asked aloud, "Where did you take me?"

"You are in my dimension now," she replied.

Still baffled, I asked, "But I thought your dimension is lower than mine, so why am I here?"

She paused to think and an embarrassed look came across her face; I could tell that she didn't have an answer, or at least an answer that I would understand. She muttered, "Well, I don't know," and quickly looked away.

I stood there thinking and remembered that I had gone to sleep in a negative and exhausted mind state. There was the answer to my question right there! I decided to lighten the conversation up a bit and exclaimed, "This is very strange! This is my room, but not really my room."

At this she asked cheerfully, "Come! Do you want me to show you around?" I nodded and followed her for a tour around my own apartment building. I could not help but notice again how familiar yet strange it all appeared. It had the same shape and layout, but a drastically different composition. I told her about this and asked whether, on the previous night, I was in her dimension or she was in mine.

"I don't know, but sometimes I go to your dimension and it takes a lot of energy. It drains me. I've actually seen you around a few times before, and yesterday I finally had the courage to approach you."

The pieces were starting to come together. Numerous times prior I had felt the presence of somebody or something, and I would quickly snap photographs that, when developed, would only show shadows or orbs. It all made sense: I was being visited by earthbounds from lower dimensional existences who were using tremendous amounts of energy in hopes of making their presence known to me.

She broke the silence: "I go to lectures, too, like you." I looked at her with a confused look on my face. "No, not in your world," she continued. "We have two professors who died, one in 1991 and the other just recently. One of them used to be a chemistry professor and the other taught history, but now they both just lecture on earthbound realms," she explained.

I looked at her in disbelief and asked, "But why? Why would they stay in a lower dimension rather than going to a higher one, if they hadn't died with negative emotions or in a negative state of mind?"

She explained to me that in the case of the professors, they wanted to further understand the connections between their dimension and the physical one. They had chosen to stay because they thought it was part of their lesson. "You know, they think that both Einstein and Tesla were aware of multidimensional realities."

"But what about you? Why are you still here?" I asked.

She looked at me and hesitated for a few seconds, as though our conversation had entered uncharted territory. But she must have sensed my dire curiosity as she continued, "Well, when I died I saw a white light appear, and instead of approaching it I ran the opposite way since I didn't want to leave my twin sister. I wanted so much to see what would happen to her and look after her. After that, I met friends, a group of people, and we've become a community over time. The bonus of being here is that we get to play pranks on the living and people we know. But I think that the dimensional securities will probably come for me soon. They never force us . . . so they tend to know when we're ready or when we want to go or not."

Upon hearing her response, I knew I had a lot to share with her, too. I told her that she would still be able to check on her loved ones and see them even from a higher existence. But she did not believe this, telling me that in a higher existence, one is only be able to see one's loved ones during the first week of death. I told her that this wasn't true, and described how I'd seen and visited people I knew who had died.

"But you're different," she responded. "You know this is real and you're aware of the multidimensional existences. But imagine if I go to a higher existence only to be able to come to my sister in her dreamworld... She'll just brush it off as though it's a dream full of subconscious gibberish!"

"So right now, your sister acknowledges your presence?"

"Yes, she knows I'm around. I play pranks on her all the time. She needs me."

I shrugged and said, "Okay, the choice is yours. You've had many lives, probably loved many others besides your sister, and some of them are possibly roaming in the higher existence right now. As you know, this universe functions based on the law of free will. It's your choice to leave to a higher place or not. There are many more things to this existence than what we know; it's not just the human world and the human cycle of living. There are many different types of planets and beings. It's an exciting world! If I were you, I'd leave in an instant. I've seen the higher existences and other planets; I wish I could be free of this human world and simply roam. Because to me, freedom is the most important feeling to indulge in—the freedom to think, the freedom to roam, the freedom to be free of the encased human mind of materiality and superficiality. Everything relating to the human world is created so that humans don't break free from their own indoctrinated minds. The worlds above are far, far, *far* better!"

I could tell that my words had sparked interest within her, as I heard her chuckle and ask, "You really think so? You make it sound so wonderful!"

I replied, "Of course I think so! Do you know the most wonderful feeling I've ever felt while traveling multidimensionally?" She looked at me with a curious gaze and I continued, "Flying—the freedom that comes when you fly and roam to different galaxies! There's a famous Persian poet by the name of Rumi, who was well aware of the truth of the human mind's capability. He once wrote, 'You were born with wings. Why prefer to crawl through life?' It's true—humans crawl... walk... even create humongous, heavy, costly machines to go to Mars, when the key is within their own minds! And you, my friend, have graduated from your lessons, so why would you want to stay in the same grade?"

I recall her staring at me with her big blue eyes, smiling as she said, "You're right, I never thought about it this way before. So, what do I do now? Should I wait, then, until the dimensional helpers come for someone, and tell them I want to go, too?"

I shook my head and told her, "Silly! You already said it earlier! When you're ready and willing, they'll know to come for you. So you have to tell them, request them to come for you; tell them the time is now. And, of course, remember to fill yourself with love. No matter how cheesy it sounds, it's true! The higher sphere is made up of love, and love itself is the key for you to go up there. Go on—try it!"

I remember her enthusiastically reciting the request out loud, calling upon the higher beings, as I did the same for her. Suddenly, a portal-like tunnel of white light appeared before us, rather than the spherical formation I had previously seen in my travels. Using both my mind and heart, I gave her all my love and encouraged her to proceed. I saw her go into it and when I attempted to follow her, I failed and once again found myself thrown back into my physical body with incredible force.

Everything I said to her was based on knowledge I had learned from my travels and adventures. But most importantly, it had all come from my heart. The worlds above are indeed *far* better—and less distant—than we think.

The Story of a Mother

Accidental journeys occur when one's physical body is in a state of exhaustion, and this exhaustion often sends the unprepared traveler to a lower-dimensional realm. Such was the case in the journey that I will discuss next. At the time I had not had a decent night's sleep in days, and my state of being was in a lower vibrational frequency due to the sadness I was feeling. I did not expect to travel energetically that night, thinking instead that I would merely dream, but the universe proved to have other plans for me. I later realized that this particular experience was meant to teach me more about the lower-dimensional realms.

As I was fading into slumber, I suddenly felt as though I were sinking into my bed and being sucked into it by a strong, invisible magnetic force. I found myself in a state of nothingness in which everything was pitch dark and I couldn't see anything. I heard what sounded like the low tones of piano music playing in the background. As all well-versed travelers know, the music you hear during an energetic journey will often resonate with the collective vibrational spectrum of the beings that reside within that particular dimension. For example, beautiful, high-frequency music indicates a higher-dimensional field in which love and light are the collective forces, while with low-frequency music the opposite is true.

I had landed in a place of complete darkness and low-vibrational music—a lower-dimensional realm—so immediately I was on my guard. I heard a man laughing at me in a sinister, mocking manner. I watched as he and a group of others approached me, and from my intuition and past experiences I could tell that they were negative beings that intended to scare me. I felt a deep fear and tried to remain on guard. As they came closer to me, their snickering was interrupted by a woman with an American accent, who sternly ordered the negative man, "Leave her alone, Jake!" They left the area with an air of bitterness, and the woman who had saved me approached me, saying, "You are obviously not from around here, are you?" I told her that I wasn't, even though it must have been obvious.

She noticed my confusion and explained, "You are currently in a dimension where the majority of people who have committed crimes, committed suicide, or experienced traumatic deaths go." She proceeded to explain to me that this dimension provides a controlled habitat for those who carry within them a plethora of negative emotions, such as hatred and revenge. These emotions are virtually engraved within these individuals. Surprisingly, most of them actually thrive on the sadness in their souls. She told me that their existence is in the *interdimension*, the dimension located between the physical dimension, where we live, and the lower dimension, where the lower entities reside. At this point, I understood that if these beings continued to roam unchecked in this interdimension, over time they would eventually become lower-dimensional entities with gruesome features and appearances.

She continued and told me her own story. She explained that she was from California, and that the only love and happiness in her life had been her young daughter. Horrifyingly, her daughter was brutally murdered, and the killer had covered up his crime with what looked like an accidental fire, enabling him to get away with it (as is the case with most crimes, unfortunately). After the murder of her beloved daughter, she dedicated her life to exacting revenge on the murderer, but one day, after having had enough of this life of anger and vengefulness, she decided to kill herself. She told me the murderer's name and begged for me to report the name to the police once I returned to my physical body so that she could finally rest in peace, as she explained to me. As I had with Pak Tejo, I simply had to refuse; I knew that death in an excessively negative state of mind often leads to existence in a much lower-dimensional realm. I immediately knew that if this woman continued to live with this dark hatred ingrained within her and the hunger for revenge eating away at her soul, she would soon begin to vibrate in the same frequency as the frightening lower-dimensional entities. Nevertheless, from my many encounters with earthbounds, I was certain that she was not a hopeless case and that I could reignite the light within her.

She guardedly thought about what I said, as I explained to her that all things in life are temporary and cyclical, composed of both an ending and a beginning. I explained to her that everything that had occurred during her time on earth was a mere lesson, and that her daughter was in another state of existence. I told her that if she wished to meet the spirit of her daughter, she must elevate her frequency. If she did so, I knew that she would come back to earth to live another life. I begged her to remember her daughter's love, her one and only life-force, and told her that this was the only way to escape the dimension in which she was trapped. I also told her that we could manifest a portal through which she could travel to the higher dimension, as long as she promised me that she would cease dwelling in negativity and hatred. She promised me, and I watched her go through the portal we had manifested, something I had learned by that time that anyone was capable of doing. After this, I was abruptly pulled back into my physical body. After waking up and making myself a cup of hot tea, I began to feel icy chills and a cold sweat overcome my body. I could sense my ghost's presence—the woman briefly visiting me on her way to the higher dimension.

Through my multidimensional travels, I've learned that love is the greatest energetic force of all and the singular tool for universal communication in the higher realms. The quote "love makes the world go 'round" can be interpreted in many different ways. Obviously "round" is the shape of a circle, which has no beginning and no end. In the context of energetics, it is a symbol of the eternal force behind *all* cyclical existences of the universe—the force of love. This is the reason why, in various ancient scriptures, the circle is depicted as the most powerful and *whole* geometric expression on planet earth. Which it is.

The Musical Sphere of the Earthbounds

In my experiences, traveling to lower dimensions, with a few exceptions, tends to involve interactions with earthbound spirits who are trapped in these dimensions and filled with strong emotions of hatred and a desire for revenge. For this very reason, I always avoid traveling here unless I am being pulled into these lower dimensions for a certain purpose. However, this next particular fated journey of mine proved to me that love does exist in this "interdimension," and that therefore the souls in these dimensions have not been entirely overcome with negativity and hatred.

Once again I arrived into pitch darkness where I couldn't see a thing. This time I heard a deep man's voice speaking to me: "Why don't you sit down over here and witness our only form of entertainment in this interdimensional level, where darkness is the only thing that exists?" The "entertainment" soon began, and I heard a woman who sounded to be in her mid-20s, her melodious voice singing and vibrating through space with her love and sorrow. She was singing of the life she had led before and how she had ended it. She sang of the strong love for her husband that resided within her and of how he had betrayed her, combining feelings of deep sadness and pain with those of enduring love. I learned that she had killed her husband as revenge for his betrayal. A man's song followed her own, reminiscing of the one great love he had in his life before it was ended and how he had destroyed it by cheating. He sang out of desperation, begging for his love's forgiveness in the chorus of the song. It was at this point that I realized that this man was, in fact, the lover of the woman who had sang before him. The third performer was a man in his late 50s. He had an extremely thick Southern accent and sang nostalgically of his previous life, dwelling on how he had taken it for granted and dismissed it as unsatisfying. He continued, singing of how he had realized, once he had entered the interdimension, that his home was in fact extremely vibrant and full of love. His song was titled "Grey City," and the chorus went something like this: "Grey city, how lovely it is, sipping a cup of hot tea. Grey city, how I miss thee. Polluted by beautiful."

After he ended his song, I heard a lilting voice rush through my mind, singing, *Music is best heard in the dark.* I was also told that various musicians over time have been "inspired" through subconscious travels to the differing realms, and upon waking up found themselves writing an ingenious piece of music. It became clear to me that these performers wanted me to bring their music to the physical world, for they were living in complete darkness and wanted to be heard.

When I returned to my physical body, I remembered this experience in great vividness. The absence of sight enables us to fully appreciate and hear music in its every detail, for it is in this state that this dominant sense is closed and the brain remains solely focused on sound. I discovered that one of the most valuable uses of music is to tell the story of our life through it. True music is a form of collective expression in which everything is conjoined. Thus, you can appreciate another's story merely by listening and connecting yourself to the words and vibrations. Music is an invisible force that connects everything within the universe. The right kind of music can certainly elevate one's frequency, while the wrong kind can lower it and block the listener's true potential.

Helpful Tips

For beginners

1. **Keep an open mind**. It is impossible to travel multi-dimensionally when attempting to do so with a closed mind. The mind is the key to unleashing the limitless wonders of the universe, and it is within the mind that the consciousness resides. Thus, if you entertain or hang onto biased thoughts that are based on what you have been educated to think is impossible, you are in actuality imprisoning your mind and depriving it of its universal right of freedom. It is of utmost importance to realize that although your spiritual mind resides within your physical brain, it is not bound by it, and that it is in fact part of your spiritual journey to eventually be able to break free of this physical shell.

   The bottom line is that when you are traveling multi-dimensionally, you are going against a universally accepted understanding of the concept of time and space. My advice to you is to forget all your pedagogically learned conceptions of the world. Open your mind and believe in the limitless possibilities!

2. **Never fear.** Fear always blocks us from our true potential! It is a catastrophic disease that plagues the human mind. Freeing yourself from fear is a necessity if you wish to travel energetically. As the wise ones of the past have stated, fear is simply the lack of knowledge. Therefore it can be eliminated through exposing ourselves to knowledge. Do not fear, for if you fear while attempting to travel, you will either be sent back abruptly into your physical body or entirely blocked from traveling. Fear is

indeed a challenging hurdle, and learning to circumvent it is a necessity for traveling multidimensionally.

3. **Don't take drugs.** I am entirely against the use of drugs for both recreational and spiritual purposes. In our modern society, in which we are continuously being brainwashed by the mainstream media, a great amount of self-discipline is needed to not fall victim to the ideas that they disseminate. Many young people turn to drugs in order to gain acceptance from their peers. Unbeknown to them, doing drugs actually lowers their overall vibrational frequency and alters their thought patterns, making them susceptible to negative energetic presences, and creating an extremely dangerous and unnatural misalignment of their energetic bodies. I'll talk more about the effects of drugs in Chapter 5.

4. **Discipline your mind.** The mind, as I have previously explained, is the key to traveling energetically. Disciplining the mind is of great importance, as our thoughts directly affect our energetic state. For example, if you are someone who tends to ruminate on negative thoughts, your energetic body will probably be a murky or unpleasant color. And, as I've already mentioned, a low-vibrational frequency makes multidimensional traveling difficult and even dangerous. With this in mind, it is crucial that you learn to control your thoughts. The society we live in is virtually devoid of spiritual truth; therefore, it is of no surprise that many of us tend to dwell uncontrollably on unpleasant thoughts. But worry not; there are various protocols that you can use to control your thoughts. Meditation is the most effective method for clearing yourself of unwanted thoughts and negative energies. Quite like an untrained animal, your mind must be tamed and disciplined over time!

5. **Keep your emotions under control, too.** Ascetics of the past were well-informed on the energetic reality of all beings. They understood that an immense amount of energy is required to travel energetically at will. Your emotions and your energetic state of being are interrelated, and your energy will dissipate if you do not have your emotions under control. For example, when you are uncontrollably angry, you are literally leaking energy. Having complete control of your emotional life is extremely important.

   Energy leakage is also likely in crowded places. Energy follows the simple rule of diffusion in biology, which states that molecules move from a greater to a lesser state of concentration. So someone who meditates daily and maintains a positive state of mind possesses a great deal of energy, and would therefore likely find his or her energy susceptible to being transmitted to someone with a low amount of energy. The exhaustion that this person would feel is the result of energy leakage. *Energetic diffusion* can be prevented through the sheer will of the mind, and the will of the mind is generated and maintained through the constant and consistent use of words of empowerment that remind us that we are masters of our own thoughts. Words are incredibly powerful, as they possess their own unique vibrational frequencies. By using words of empowerment, you can actually teach your mind to automatically govern your energetic state. Use the power of words in order to keep yourself energetically composed.

6. **Avoid alcohol.** It is inadvisable (and actually dangerous) to travel energetically under the influence. Being absolutely conscious and in full possession of your faculties is necessary for positive multidimensional experiences. Traveling energetically under the influence is similar

to driving a car in the same state: It disorients you and weakens your ability to react in a timely manner.

7. **Embrace the power of intention.** Intention plays a decisive role in traveling multidimensionally. The power of intention is strengthened through positive thoughts and the pure willingness to be of service to all. In order for you to embark on a journey that will unleash the greatest knowledge of the universe, you must make it your intention to be of service to humanity. In short, you must not desire to travel multidimensionally for selfish reasons; instead, be entirely willing to share what you learn from your adventures in a non-egoistical, selfless manner.

Steps for helping earthbounds

✳ **Acknowledge their presence.** Earthbounds live their lives invisible from the rest of the world. Many cause mischief and even menace the living in an attempt to draw attention to themselves. Thus, simply acknowledging their presence often satisfies them.

✳ **Teach them to let go.** Most earthbounds have died in a negative state of mind, the majority of these with unfinished business. It is important that you show interest in their life story, as in many cases they are stuck in a state of limbo and wish to simply share their story with a living person. By speaking to earthbounds who continue to dwell on their past, you are teaching them how to let go. It is crucial that earthbounds learn how to do this, as it is impossible for them to move on and ascend to a higher state without first letting go of their past.

✳ **Inform them of the cyclical nature of existence.** Many earthbounds die forgetting that the traumatic life they led is only one of many they've had. Most of them are even unaware of the existence of the higher realms and remain trapped in a negative mind state, which prevents them from ascending to the higher-vibrational existences. Therefore, it is important that you inform them of the cyclical nature of human existence, and encourage them to look ahead instead of dwelling on their current state of being.

✴ **Help them visualize a better existence of positivity and love, through collective manifestation.** Love is the ultimate antidote for raising one's vibrational frequency. Consequently, in order to successfully assist earthbounds in their ascension to a higher realm, we must first raise their vibrational frequency, and this can only be done through love. The majority of earthbounds, though trapped in their negativity, are not entirely devoid of love. In fact, with our assistance most of them are able to once again fully comprehend the power of true and pure love. Consequently, we must remind them of the love they have (or once had) for their loved ones. Our minds possess the incredible power of manifestation, and, together with an earthbound, we can envision a much better, higher existence for them. To do this, we use the power of attraction to send our thought waves to be picked up by the higher-vibrational realms. The power of manifestation is the ultimate solution for assisting an earthbound into a higher-dimensional existence.

Throughout the ages, various religions have used this method of manifestation to assist their deceased. Interestingly, it is with this knowledge that Buddhist *sutras* are continuously recited at funerals. Take, for example, the *Amitabha Sutra*, in which a higher realm is portrayed descriptively through the use of poetic incantations. It is these particular recitations that will assist a newly deceased Buddhist in using the power of manifestation. It is generated from automated imagination and triggered by the illustrative depictions of the heavenly scenes beautifully portrayed in the sutras.

Now that you have a handle on some of the basics of multidimensional travel, we'll discuss some of the more salient facts and implications of this ability a bit more in depth.

2

Soul Connection

Adeep soul connection is a connection that persists through time and space. It is one that is initiated, strengthened, and nurtured through many lifetimes. It is a strictly positive connection in that it lacks both negativity and karma. Souls sharing such a connection tend to meet each other through their lifetimes as a means of supporting one another through arduous and challenging journeys. The love that exists between two souls sharing such a connection is one that is often described as "unconditional." Speaking more personally, I am certain that a deep soul connection existed—and still exists—between my grandfather and me. Though he was unable to fully comprehend the spiritual ramifications of my experiences, he always offered me assistance in a perfect synchronicity of timing. His assistance to me was what many people would describe as heaven-sent or providential.

In the past, I often puzzled over how he could always sense how I was feeling and then offer me help at exactly the right time. It was only after his passing that I solved the puzzle: that indeed a soul connection existed (and still exists) between us. This soul connection gave us the ability to communicate telepathically via intuitive sensing, a phenomenon known as *soul telepathy*. Telepathic ability is not uncommon, and it can be sharpened and perfected over time. However, the strongest telepathic connection is one that exists between two souls who share a soul connection.

Grandfather's First Visit

In the spring of 2013, I traveled around Japan for three weeks with my dearest friend, Mary. She had originally expected our trip to consist of simply shopping and eating, but little did she know, I had planned all along to transform our trip into a spiritual one. We traveled together around the vast island of Japan in search of archeological areas of interest and evidence of past civilizations. We embarked on our adventures with boldness, visiting abandoned homes in search of evidence and documents to secretly sneaking into hidden temples containing mummies similar to the ones found in Egypt. Though our adventures were spiritually oriented, I could not help but feel rather restless throughout the entire trip, as though I were meant to be someplace else. I was baffled by the uneasy feeling in my heart, and it was only during the second week of our trip that the meaning of this feeling became clear.

While we were in Fukuoka, my grandfather, who was alive and well at the time, visited me energetically for the first time. I found him sitting at the edge of my bed, calmly informing me that it was time for him to go, that death had finally come for him and I ought to take good care of myself. His words brought me great alarm, and I selfishly begged him to not leave yet. I recall seeing his energetic body dissipate slowly and I immediately forced myself to wake up. As soon as I found myself back in my physical body, I worriedly reached for the phone next to my bed to call my father and ask him about my grandfather's health and whereabouts. My father told me that my grandfather was doing all right and at home, but that he had experienced constricted breathing the day before and hence had to use an oxygen mask to assist him with his breathing. Despite his reassuring reply, I was overwhelmed with feelings of uneasiness for the rest of the day, and my intuition began nudging me to leave

Japan as soon as possible to see my grandfather. Nevertheless, I felt that I couldn't leave my friend Mary all alone, so instead I used the power of logic to force myself to believe that my grandfather was well.

I stayed in Japan for another six days, unable to enjoy our many hikes, Shinkansen trips, and archeological adventures, remaining constantly worried about my grandfather's well-being. During the entire trip, I wondered how he had managed to travel to me that night; I thought that perhaps the will within his mind had been strong enough to allow him to detach from his body in order to relay the message to me. I was well aware that the phenomenon of detaching from one's physical body is entirely possible when the strong will within one's mind overcomes the physical senses. Ironically, my grandfather was a true old-fashioned conservative who had no interest in spirituality whatsoever, so his visit naturally baffled me.

Six days after my grandfather had visited me, as I was eating dinner in my hotel room, I received a panicked phone call from my grief-stricken family. They informed me that my grandfather was in the emergency room, basically fighting for his life. Apparently, he had been hospitalized four days earlier, and, attempting to not ruin my trip and assuming that his condition was not very serious, they had chosen to keep this news from me. I recall being incredibly disappointed in myself; it was this particular incident that taught me one of the greatest lessons of my life—the lesson to always choose intuition over logic. Logic, which is a child of the five physical senses, tends to be faulty when compared to intuition, which arises from the invisible and largely misunderstood metaphysical senses and supernatural abilities.

A Death Observed

The moment I received the sudden news of my grandfather's critical condition, I quickly purchased a ticked for the next flight home. Unfortunately, the flight alone lasted 22 hours, and when I arrived, he was mostly unconscious and had been placed on respiratory support. The doctors had seemingly turned to desperate measures and decided to insert a breathing tube into his mouth, and consequently he was left unable to audibly convey his last words to me.

In the rare moments that he was conscious, I would gratefully gaze into his eyes, mouth my words, and tell him not to speak, while he would reply to me non-verbally through his eyes and tears, as though he was saying good bye. Seeing my grandfather, the same man who refused to ever show any feelings of pain or vulnerability, in such a state upset me terribly and only served to increase the disappointment I felt in myself. I could not keep myself from thinking that if I had only followed my intuition in Japan and flown to him immediately, my grandfather would have been able to convey his last words to me through speech. These thoughts ran rampant through my mind and pulled me into a state of utter hopelessness.

After my grandfather developed a lung infection, it became dreadfully clear to me that I would never hear him speak again, and my prediction proved to be unfortunately accurate a week later, when the doctors declared that he must undergo a tracheostomy. When the doctors broke this news to us, I remember gazing at him and noting that he had been in REM sleep for most of the time. And it was at this exact moment that I forced myself to snap out of my depressed state and to use my knowledge of multidimensional travel to assist my grandfather. In order to travel that very night, I used my basic protocols for traveling multidimensionally (see the Appendix for a quick and easy reference guide to these protocols) and managed to maintain a clear mind. I also memorized the route from my house to the hospital. Many books suggest that during out-of-body travel, one becomes a much more intelligent version of one's self, but this is a highly misinformed notion. In actuality, when traveling out of body, you are *entirely* yourself and retain the same memory and thought patterns that you would have in your physical body. Thus, being practically useless without my navigation system in the physical world, I knew I had to memorize the directions to my grandfather's place in order to travel there successfully.

That night, after much dedicated preparation, I managed to travel to my grandfather. I found him in a half-conscious state; in the area where his energetic body met his physical body, I noticed a dark spot on his stomach region. As I was examining it, I was shocked to receive a strong feeling that there was a tumor in or around his stomach. Almost in an instant, I rushed back to my physical body and wrote down what I had discovered, trying hard to remain positive. I had already learned my lesson to always pick intuition over logic, so I was desperate not to make the same mistake again. It's worth noting here that if your emotions and thoughts are not positive and your subconscious is not tamed after an out-of-body journey, you will retain little to no memory of the experience upon returning to your physical body.

The following morning, I forced myself out of bed at 5 a.m., precisely three hours before my grandfather's scheduled tracheostomy, so that I could pray to the universe for a miracle to delay his operation. Exactly one hour later, my prayer was answered. We received a phone call from the hospital informing us that the doctor in charge of my grandfather's operation had advised us to delay his operation by two days.

I spent the following two days trying to convince the doctors to perform an ultraviolet scan on my grandfather to check for a tumor. My attempts proved to be futile, and much to my dismay and discouragement I was pushed aside by the doctors who bluntly stated that the possibility of a tumor was nonexistent and they would have been aware of it if my grandfather had any tumors. Usually a collected and reserved person, I grew increasingly frustrated and even resorted to shouting for an ultraviolet scan. Nevertheless, everyone simply turned away with a deaf ear, and on the third day the doctors performed the tracheostomy as scheduled.

Fortunately my grandfather survived the operation, but his health deteriorated exponentially following the procedure. His entire physiological system began to collapse simultaneously—his kidney malfunctioning first, followed by his bladder, and then his liver. It was only after his liver failed that the doctors decided to go through with an ultraviolet scan, and to their astonishment (and my sorrow), they detected a large tumor in my grandfather's stomach region. By this stage, though, it was already impossible for them to remove it.

Following the doctors' discovery of the tumor, I fell further into emotional despair. Due to my negative state of being, I was unable to both sleep and travel multidimensionally. While attempting to sleep for the first time in what seemed like days, I heard a man's voice communicate through my mind: "Two weeks. You've been asking. It's two weeks—12 a.m.," giving me a concise answer to the question I had been seeking in my mind, where it was open and free to be voluntarily answered by beings of the various dimensional existences. And indeed, 12 days dragged by painfully and my grandfather saw no improvement to his health.

At that time I had resorted to coming to the hospital in the afternoons instead of the mornings, allowing me to begin my day with meditation. On one of these mornings, when I was absorbed in my meditation, I received an unexpected phone call from my delighted mother, ordering me to quickly head to the hospital. According to her, my grandfather was fully conscious and had improved tremendously, "by some sort of miracle." While running through the depressing hospital hallway toward my grandfather's room, I spotted my mother smiling widely as she exclaimed, "Glowing, glowing! He's glowing!"

Sure enough, when I stepped into his room, I saw him sitting upright on his bed, his face glowing with liveliness and vigor. It was as though he had received a tremendous boost of life-force from a higher being! He smiled warmly upon seeing me and stared directly into my eyes. He was still unable to converse with me through words, but I exclaimed to him happily, "Grandpa, you have to be well! I'll take you back home and we can eat ice cream together with Grandma, like we used to!" He simply continued to grin and look into my eyes, and I felt that he was telling me to take good care of myself. "Grandpa, you don't need to worry about me. I can go wherever you go. And remember, you made me a promise last year that wherever you go, you will surely keep in contact with me. Grandpa, remember: I promised you last year that I would begin writing a book? I still remember that promise, Grandpa. I'll dedicate the book to you. That's why you can't leave just yet. Grandpa, don't you want to see my book?" He responded to me with his usual humble, loving smile. "You've been asleep most of the time. Did you see your grandma whom you loved so much?" He nodded slightly. But snapping myself out of this state and not wanting to let him go, I suddenly burst out, "Grandpa, your health has improved so much! Even the doctors said so—you can't lose hope! We'll go back home tomorrow and we'll eat ice cream together with Grandma. We all miss you so much!" Our joyous yet bittersweet "conversation" was soon cut short by the nurses and I was asked to leave, as it was time for them to clean him up.

Afterward, my family and I had a wonderful dinner together at the hospital, rejoicing, joking, and laughing for the first time in the days since my grandfather had fallen severely ill. I left the hospital around midnight feeling incredibly relieved and happy, and upon returning home I decided to take a relaxing shower. Almost abruptly, I was interrupted by loud heavy knocks on my bathroom door, followed by my mother's frenzied voice instructing me to get ready and get into the car, *now*. We were going back to the hospital. On the way there, my mother morosely announced that my grandfather was in critical condition and had gone into a coma; according to the doctors, he could pass away at any minute.

When we arrived at the hospital, we found my grandfather unresponsive and in a coma. In an energetic sense, a coma is a state in which the consciousness has entirely or almost entirely departed from the physical body. With this in mind, and because I knew that the last bit of my grandfather's consciousness was being transferred out of his physical body, I knew I had no choice but to meditate in order to communicate with him. Even though I had previously been embarrassed about doing this in front of my closed-minded family members, I dove into a meditative state and immediately sensed a presence in the room. Focusing on it, I received a message: *An individual's destiny to return to the spiritual realm should be celebrated.* The voice resonated clearly through my mind, but not through my ears—a telepathic communication. I soon recognized that it was the same inner voice that had previously informed me of the timing of my grandfather's death. Maintaining this communication, I replied telepathically, "I know it is, but please give him more time—just a little bit of time? I beg of you. I am not ready to let him go. Please!" As my thoughts transversed the borders of my mind into his, the hospital room was overcome with embraces and sighs of relief. To my bewilderment, it was as though my grandfather had been sent right back into his physical body at that moment, for his eyes opened wide and he seemed to be fighting to breathe. I broke out of my meditative state as quickly as I could and embraced him tightly. Later that morning, we all returned home feeling relieved. I felt incredibly content and thankful for the higher beings, and I was too tired at that point to even care what my narrow-minded and conservative (yet loving) family thought of me.

I had not planned to travel consciously out of my body that night, so I was startled and overjoyed to find my grandfather standing next to my bed. Noticing my smile, he said calmly, "I have to go."

His words deeply saddened me, and I cried out selfishly, "You can't go! What's going to happen to me if you go?"

He looked at me, hesitating a bit, but then repeated slowly in a serious tone, "I have to go; you have to let me go."

I told him again that he couldn't. There was a brief moment of silence between us, before he said lovingly, "You have to let me go. What else am I supposed to do? I wouldn't be able to live normally. My legs are useless, both of my arms are tied, and if I live, I wouldn't even be able to eat or speak. I'd be a vegetable. Imagine if you had to live your life in such a state... You have to let me go. And remember, I want you to take good care of yourself. I'm sorry that I won't be able to take care of you anymore. You have to be able to live your life independently. Remember to take good care of yourself, always."

At breakfast the next morning, shattering the general happy mood, my father announced to the family that my grandfather was back in a coma. We continued our meal in silence until my father turned to me and sternly said that I would have to let my grandfather go that day. My other family members chimed in and agreed, commenting on how incredibly selfish it was of me to hold my grandfather back from his death. As I listened to their chatter, I grew exceedingly upset at both their remarks and the thought of having to let my dear grandfather go, but I began to wonder if perhaps they were right. I remembered my grandfather's visit to me from the night before, and after envisioning him in the continuous vegetative state that he had described in our conversation, I quickly came to my senses and vowed to let him go that day. Above everything, I wanted my grandfather to be fully prepared before he passed into spirit.

That day, I arrived at the hospital earlier than usual in order to convey to him the basic knowledge he needed to navigate successfully through the energetic dimensional realms. I stood beside his bedside, holding his hand so that it would be easier for him to align himself to my frequency of thoughts, and began teaching him the basic concepts. "Grandpa, I have three messages that you must remember. Firstly, in the world that you're going to, the mind is *everything*. Therefore, you need to control your mind. You can't think of anything or anyone negatively anymore. Let go of everything and envision the higher existences that people know as heaven. Think of that place; envision it within your mind, Grandpa. Secondly, you cannot feel any sadness or anger. Instead, you must only feel the love that you have for me and all of us, and only feel that. Trust me, Grandpa. Third of all, I know you dislike asking for help, but in the place that you're going to, nobody will guide you or help you unless you ask them to. So ask for assistance to go to a higher existence, and they will come for you. You'll soon discover that there exist many dimensional existences apart from the one that we live in, and I want you to go to a much higher one. You've graduated from your journey with flying colors. Other people having to go through what you did would have given up a long time ago, but you passed with flying colors! You have nothing to be afraid of; let go of everything! I am letting you go, okay, Grandpa? I love you, so I am letting you go. You can go now. We will meet again, I know it. I'll be there in approximately 40 years, earth time. Thank you for being my greatest pillar of support, Grandpa. I am very blessed to have met you again in this life. Thank you for the unconditional love that you have given me. Go in peace, Grandpa."

I stayed next to him holding his hand tightly and guided him in his passing into spirit for the next 10 hours, leaving only to get a drink of water and use the restroom. My grandfather passed away at exactly 12:03 a.m. Although I had the knowledge of life after death to console me, I broke down and cried on the cold floor of the hospital room after his passing. To this day, it is my grandfather's great love that has served as the key to my multidimensional travels to the higher realms. And, it is thanks to his immeasurable love that this book was written.

The day after my grandfather's death, I arrived at my grandmother's house in a dilemma over whether or not I should tell her of my grandfather's passing. To my great astonishment, I found her sitting on the bed staring at a photograph of my grandfather that was hanging on the opposite wall; according to her nurse, she had been in this exact position all morning. My grandmother had unfortunately already entered the later stages of Alzheimer's disease; most of the time she was unable to remember her own name, let alone the fact that my grandfather even existed. As she continued to longingly stare at the photograph, I approached her slowly and asked her how she was feeling. With her eyes still glued to the photograph, she replied sorrowfully, "My beloved came to me last night to say goodbye. He told me that he doesn't want anything anymore, that he was going to leave."

Having expected her usual incomprehensible response, I looked at her in complete shock and grief before answering, "Well, you love him, don't you? It doesn't matter if he's gone. Love is everlasting."

Upon hearing my words, she began to sob loudly and, grabbing my arms demandingly, asked in a distressed tone, "How can I love him if he's gone?"

I sat down next to her and began to rub her back, attempting to comfort and calm her with my soothing words. Within minutes, however, she seemed to have forgotten everything she had just told me and asked for her breakfast.

A Funeral

Although my grandfather had never truly followed any religion, and in fact seemed to follow his own religion of love itself, he was raised as a Buddhist. Because of this, his funeral was set to follow the style of a Buddhist ritual and procession. Prior to his funeral, the monks that my father had hired asked that we place my grandfather's bones in a jar in order to be eternally prayed upon in their temple. I was in a state of utter disbelief over the monks' request, for I had previously learned from my many multidimensional encounters with people who had died, that this is possibly the most atrocious way to treat the deceased's physical body. Because bones retain small aspects of the vibrational energy of their owner, such a treatment would be a disturbance to the owner's peace. Therefore, I told my father that I had only two requests: firstly, that my grandfather's bones should be properly disposed of into the ocean; and secondly, that his body be cremated only after a minimum of three days had passed, as the energetic body would still be connected to the physical body during that time. If cremation is done too soon, the deceased will sometimes feel the effects of what is being inflicted upon the physical body. I additionally requested that my grandfather's body be disposed of in water due to its potent purifying nature. Much to my dismay, my father, himself struggling with his grief, replied condescendingly,

"Who are you to question the monks? They've spent their entire lives studying spirituality, but you know nothing!" In that moment, I recalled my grandfather telling me once that it is impossible to teach the unwilling, so I remained silent.

On the day before the cremation of my grandfather's body, I spent the entire day quietly meditating in the funerary resting room. The room was located in a secluded area at the back of the main room, where my grandfather's coffin had been placed and incense and candles were lit. With the warm scent of the incense and my midday weariness pressing on me, I had to remind myself several times to not fall asleep, which proved to be quite difficult, thanks to the large collection of comfortable couches and soft pillows the area offered its visitors. Nevertheless, I ended up dozing off to sleep, and was able to naturally detach from my physical body.

Almost instantly, I found my grandfather in the main room, shuffling back and forth impatiently beside his own coffin. He looked exactly the same as when I had last seen him in the hospital, no younger or livelier, and still bearing the same open wounds that he had acquired from his operations, such as the incision on his neck. His face lit up immediately upon noticing me, and he asked with enthusiasm, "Is it really true that my legs can't be fixed?"

Wanting to be entirely honest with my grandfather, I replied, "No, Grandpa."

"That is a shame! I still have a lot of things that I could do in this life, and would do. Are you sure it can't be fixed? Look at me! I'm walking about so fine and well. I feel good and refreshed!" he exclaimed. His behavior struck me as being strange, so I began to worry that he was perhaps stuck in a state of limbo.

"Grandpa, you're not in your physical body anymore. Your physical body is in that coffin," I explained to him in a sad, somber tone. I slowly lifted my hand to point at his coffin, struggling to hold back my tears.

He glanced at it quickly and replied with frustration, "Yes, I *know*, but it was the same when I was in the hospital. I was walking about like this normally and I could get back in. Are you really sure there's nothing that can be done? Can't they fix my legs?"

Terrifying thoughts of my grandfather ending up as an earthbound began to plague my mind, and I politely asked him to sit down. "Grandpa, your physical body is useless; it's inside that coffin. It's virtually impossible to fix. Grandpa, please listen to me," I begged in a hoarse voice on the verge of tears. "It is time for you to go; your journey here has ended. When you see the tunnel of light, which you will very soon if you haven't yet, you must go, Grandpa. Go into it."

As I continued stressing this to my grandfather, I suddenly heard the faint echo of my younger sister's voice: "Is she not waking up? Maybe she's just exhausted." "Just shake her! The second session of prayer is starting! Quick!" my cousin barked at her. I woke up to the sound of their chatter and my younger sister shaking my shoulders. She stopped only after I had promised to not fall asleep again, and the three of us headed to pray for Grandfather.

It was only after the cremation of my grandfather's body the next day that I felt he had left the physical world for good. I suspect that it was due to the infinitesimal amount of consciousness still remaining in his physical body (resulting from his attachment to the physical) that he had lingered and even considered the possibility of going back into his physical body. Interestingly, while his ashes were being released into the sea, I heard incredibly beautiful, high-frequency orchestral music resonate through my mind. Upon hearing it, I immediately knew that my grandfather had ascended to the higher sphere and that this music was the vibrational frequency of the energies of that realm. It was a sound that was familiar to me, as I had heard similar music while previously assisting others during my travels through the energetic realms. The lovely sound provided relief for my grief, and in that moment I felt entirely calm. I was reluctant to ask if others had also heard the music, too, out of fear of being ridiculed by my well-meaning but judgmental family. However, as we were driving back home from the ceremony my mother suddenly blurted out, "I heard very beautiful music playing during the releasing of the ashes. Did anyone hear it, too?"

"Your subconscious, lady. You're just too exhausted," answered my father in his usual stiff tone.

All of my family members nodded in unison at my father's comment and then directed their attention toward me, as though expecting some gesture or indication of agreement.

I lied and mumbled quickly, "Probably a radio playing in the background." But as I looked out of the car window and watched my grandfather's hometown rush past me, I felt a wave of strong, unbearable guilt. Interrupting everyone's laughter, I suddenly exclaimed, "Actually, I remember hearing it, too! That music was very beautiful. Grandpa is probably in a much better place now."

Visiting in the Afterworld

After my grandfather's death, I repeatedly attempted to travel energetically to him; however, each time I was blocked and immediately sent back to my physical body. Nevertheless, I kept trying. I knew that although it is easy to consciously visit a loved one when you are tuned into his or her unique overall vibrational frequency, it is only when the universe allows this visit that you will be able to travel to that person successfully. After my 20th attempt, I was able to finally see my grandfather again.

I felt myself flying through space with a faster-than-light sensation over what looked like a futuristic metropolitan city. Everywhere around me, I saw blue domes that looked as though they were made of glass, with a few tall and narrow buildings scattered among them. Oddly, everything was either blue or green; there were no other colors. I soon spotted my grandfather standing outside one of the blue domes, on a path next to patches of vibrant, green grass. He seemed to be expecting my arrival, and to my amazement I landed right in front of him as though I had been driven there by a powerful, unseen force.

My grandfather looked at least 40 years younger than he did the last time I saw him, and taller, too. He also had jet-black hair instead of the white hair of old age. In addition to these obvious physical differences, he seemed extremely alert and refreshed and didn't have a single wrinkle on his face. I noticed that he was dressed in light-blue attire that reminded me of the medical scrubs worn on earth. After spotting others around us dressed in the same blue clothing, I ask him out of curiosity, "Grandpa, where are you?"

He replied to me in a slightly biting tone that was unlike him, "A place where dead people roam. Where else?" I desperately wanted to embrace him, tell him how much I missed him and still cried for him, but I decided to compose myself, as I didn't want to make him sad. As he looked around us, he continued in a rather dispirited tone, "It literally is a place where dead people roam." I gazed curiously at his throat to check if the laceration from his tracheostomy was still there, but it was gone without a single trace. More people rushed past us in their blue uniforms as though they had some place to be.

"Well, I know that this is the place where dead people roam, Grandpa, but what is this place exactly? A hospital?"

He replied briskly in the same sharp tone, "Yes, hospital for the dead."

"But why are you here?"

His response sounded bitter and nothing like the grandfather I knew: "Well, it seems that I am fated to live in a hospital, either during my time with the dead or the living!"

After it became dreadfully clear to me that my grandfather had no idea where he was, I decided to stop bombarding him with questions. As I curiously studied my surroundings, I noticed a man with short, blonde hair who was dressed in a noticeably lighter colored uniform than everyone else, begin to approach us. He cheerfully addressed my grandfather by his name and turned to me and asked, "You're his granddaughter, the conscious traveler?" I nodded and he said with a warm smile, "Your grandfather just woke up today from his long, rejuvenating, and healing power-sleep! We have yet to show him around. He asked to see you the moment he was brought to full consciousness."

"Is that the reason why I've been unable to visit him?" I asked with curiosity.

He nodded reassuringly. "We've put him entirely to rest here while we worked on healing his energetic leakage. The operation he endured during his physical life was on his throat, causing great energetic leakage. All of the damage done to the physical is shown through the energetic, and vice versa."

"So this is a hospital where people who died with damage to their physical body go to?"

He nodded.

"Before this journey, I used to think that people could simply heal themselves in the energetic world and that no professional help was necessary."

He smiled at this and responded, "Well, people who generally just die aren't professionals in healing. The healing here is very different from the healing on earth; we don't perform surgery on people here. Unlike the physical body, as you probably already know, the energetic body is healed through the will of the mind. Therefore, our healers are actually adepts at restructuring the energetic body through the usage of the mind. Most doctors who reside on physical planet earth, though highly educated on the understanding of the physical body, greatly lack knowledge of the two other existing bodies. If these doctors instead sought an understanding of this and integrated these discoveries with the knowledge of the physical body that they currently possess, many lives would be drastically improved." He paused and continued thoughtfully, "I think I've given you enough information that you could take back and share, as I've been instructed to do. Your grandfather needs to go for his power rest again as he is still not fully well. Do you want to come with us to his room?" I nodded, and we entered the door and began walking through the domed building.

Inside was what appeared to be a hybrid between a hospital and hotel, but without that negative feeling I had grown accustomed to during my grandfather's stay at the earthly hospital. As we walked down a long hallway, I reminded my grandfather, "Grandpa, remember my three messages. They still pertain here. I know you belong to a much higher vibrational frequency. From the look of it, this place is just a transitional area that everyone with physical damage must go to." We soon arrived at a room with a comfortable-looking queen-sized bed, and the man instructed my grandfather to rest there. I proceeded to tuck him in and promised, "I'll come back again. Please remember my three messages, okay, Grandpa? You ought to be happy; you're in a better place than earth!" As I continued to say my goodbyes, the man suddenly turned to me and warned me that I was going to be pulled back into my physical body at any moment. He bade me farewell and, seeing that my grandfather was comfortable in bed, turned to walk out of the room. I began preparing to return to my physical body when a thought suddenly raced across my mind, and I called out after the man, "Wait! Wait! I have a question. Why does this place look so much like earth? I thought this place is in a higher dimension than earth!"

He looked almost impressed by my question and answered, "Because everyone with damage to their physical body comes here first despite their differing levels of vibrational frequencies. For this reason, we constructed it on the basis of familiarity." This immediately reminded me of the areas of the chakras, and as though he had read my mind, he nodded and said, "Goodbye, now. I will take good care of your beloved grandfather; don't worry!" I was quickly pulled back into my physical body and woke up with the calm sense of relief from finally knowing my grandfather's whereabouts. At that moment I decided that it was best to not visit him for a period of 30 days so that he could quickly adjust to his new surroundings without being distracted or disturbed by me.

On the 30th day I stared at the calendar for at least 15 minutes before making up my mind to travel to him that night. Everything in this universe exists in the perfect synchronicity of timing, and that night, it felt undoubtedly right for me to travel to him. Before going to bed, I followed the usual protocol of matching my frequency to my grandfather's and ordered my mind to go to him. I soon found myself flying, but this time, instead of landing properly, I quickly passed through a small, green dome and penetrated through many rows of walls until I ended up in what appeared to be a small classroom containing a group of students. The room had five small, round, green tables scattered throughout, and among the many faces I found my grandfather seated at a table to the far right. As soon as my grandfather saw me, the teacher standing at the front of the classroom immediately directed his attention toward me. Amusingly, he reminded me of Colonel Sanders, the founder and face of Kentucky Fried Chicken, with his iconic white hair, beard, and mustache. "Hello there! Come and join us. This is his granddaughter, and she's a conscious traveler. Please proceed to have a seat next to your grandfather."

As soon as I sat down next to my grandfather, I couldn't help but ask him what this place was. He whispered over to me excitedly as the teacher continued speaking to the class, "I, along with nine other people, have been chosen to go to a higher place. So we are required to take these classes." I pressed him for details, and he told me that he had been learning fascinating things about the history of the universe, its structural makeup, and the different realities contained within it. I then heard the teacher ask the students, "So, do you understand? This is indeed a very different world than what you have previously known. Every emitted thought is sent to the dimensional barrier through time and space. This is precisely why the concept of thought is of incredible importance."

My grandfather then introduced me to the man sitting next to him. His name was Tannuhsuhkolwit, and I repeated the name many times to myself so that I would remember it when I returned to my physical body. My new acquaintance enthusiastically expressed how excited he was to be going to a higher existence and described the classes that they were required to take as "out of this world." In the midst of our conversation I suddenly heard a loud, annoyingly familiar noise and was instantly pulled back into my physical body.

I later learned that the noise had come from yet another round of my boisterous neighbor's hardwood floor installations. Unless you are a heavy sleeper who can sleep through loud noises, disturbing noises like this one are a curse to avid multidimensional travelers. Much like an airplane that has experienced a jarring, hard landing, the energetic body is instantly sent back to the physical with the incorrect alignment, and an incorrect alignment can leave you feeling exhausted, drained, and disoriented for the rest of the day. Therefore, my biggest advice to anyone wanting to travel free from any disturbances is to choose a home with soundproof walls!

A Warm Reception

Eleven days had passed since my last visit to my grandfather. Following that visit, the management of my apartment building apologetically informed me that my neighbor's home improvements would last for another 11 days, so I abstained from traveling during that time. Instead, I had subconscious dreams. I had grown so used to traveling energetically that having normal dreams began to feel rather strange to me. Similarly, the idea of not controlling my mind and allowing it to subconsciously release images and scenarios made me feel a bit uneasy. So, when the final day of my neighbor's construction came, I celebrated and rejoiced with a sense of reclaimed freedom.

That night, I meditated briefly before going to sleep and communicated to my mind that I was going to travel to see my grandfather again. Within a second, I found myself detached entirely from my physical body and flying through space and the various dimensions. (I have always been afraid of heights, so I usually close my eyes during this part.) Instead of landing back in the transitional realm that I had last been to, I found myself in a dimly lit hallway with what appeared to be a reception area at the far end. Two youthful-looking Asian women and an old man dressed in a long, white Chinese robe came into my view as I walked toward the long desk in front of me. The man was skinny, with a long white beard and thick white eyebrows, and I noticed that he was holding an ancient-looking wooden staff. He reminded me of the wise old men depicted in ancient Chinese scrolls. As soon as he spotted me, I saw his jaw drop, and he approached me with excitement and delight evident in his face and entire bearing. He embraced me like an old friend and, as he held both of my hands, exclaimed joyously in his thick Chinese accent, "Saho! Saho! You are here! *Ni lai le!*" (You are back!) "Saho! Ni lai le!"

I was utterly confused and thought that he had somehow mistaken me for a fellow named Saho, so I said to him politely, "I am not Saho; my name is Khartika."

Despite my comment, he simply ignored me and continued, with his enthusiasm unabated, "Saho! *Wo hen xiang ni!*" (I miss you!) "Saho, *ni lai le!*" (You came!)

I repeated with some exasperation, "My name is Khartika, *not* Saho."

I saw his eyes widen in bewilderment as he asked hesitatingly, "Cart...?"

I responded hastily, "Yes, I'll spell it for you. It's K-H-A-R-T-I-K-A. This is the lobby of this place, right? Do you think that you could perhaps look me up? I'm looking for my grandpa."

To my disappointment, he avoided my question and said with a smile as he patted my back, "Saho, it is good that you are back! I was so sad when they told me you had gone back to that terrible world!"

After listening to him repeatedly express how glad he was to see me again, I asked him with curiosity, "Do we happen to have a past life together?"

He nodded enthusiastically a few times before explaining slowly and majestically: "Long, long, long time ago...when the wind was blowing hard...in a mountainous region...we had a life together in a place that is now called...Malaysia. Or is it now called Thailand? It was a long time ago...it wasn't known as that...we were studying spirituality together... and we had another life together in a place called Potala. Saho, I was your teacher! You were a good student and always have been my favorite. You are a brave soul, Saho. You have endured so much, haven't you? You went back to that awful place without telling me."

Distracted by my strong desire to see my grandfather and knowing that my time there was extremely limited, I impatiently interrupted him, "Sir, can you please tell me the story another time? I really want to find my grandfather." I realized how rude my statement must have sounded when I saw the disappointed look on his face, so I quickly added, "I am very sorry if I sound rude. If we had a life together, I would not remember it. In fact, I do not know anything about most of my past lives. I'm actually not back for good yet. I'm still attached to my physical body, in which I'm known as Khartika. I came here to look for my grandfather."

He seemed saddened by my response and asked, "You are not back for good yet? So how did you get here?" I explained to him that I did not know and that I simply wanted to find my grandfather, at which he gazed at me with concern and asked, "Who is your grandfather, Saho? Or by what name is he known as?"

My grandfather actually had three names—the first was used by his grandchildren, the second was his formal name, and the third was used by his relatives. I mentioned his first name to the old man, which garnered no response, and then mentioned the second, again receiving no response from him. It was only after I had said my grandfather's third name, the name his relatives used for him, that the old man grinned widely and exclaimed, "You're *his* granddaughter, Saho? He is a remarkable man of high virtue; he made an impression on everyone!"

I was overcome with a sense of relief and asked him excitedly, "So, you've met him? He's not at that hospital place anymore? He is indeed a great soul!"

"Yes, everyone has to pass through here before being segregated to places that match their vibrational frequency. Your grandfather made an impression on everyone! A man who came from such a small spiritual level and then was placed on such a high level...it shocked and surprised everyone here, as it is an exceptional case!" I was incredibly happy to hear this and asked the wise man if my grandfather was in a much higher existence now. The old man suddenly threw his arms in the air as he answered, "Yes, and large! Even the door to go through it is so large and wide! It's so big."

"So, it's been a while since he left that hospital place with the domes made of glass?"

The old man chuckled upon hearing me say this and exclaimed cheerfully, "Glass? It's energized crystal, not glass! Yes, he came here and stayed longer than others, as he was an exceptional case. Coming from low spiritual understanding and then going to such a big place! You should be happy for him! I spoke to him. He is a friendly and wise man." I smiled in agreement as he continued, "Saho, I thought you were back. I was not allowed to contact you. That's why I never did. I didn't even know how you were managing in that world. How are you doing there?"

Feeling compelled to give him some sort of answer, I replied, "I'm currently writing a book to help people realize the truth about multi-dimensional realities and assist them in their spiritual understanding."

"A book? But still, Saho...I do not know why you would go back. I only have one message for you: Your true nature is Saho. I like the name Saho better than that 'Cart,' too; it suits you better. If they allow me, I will contact you. I give you all of my good wishes!"

As he turned away from me, I called after him, "Wait, wait! Are you the gatekeeper here? Can I go to the place where my grandfather is? Can you open that door for me and allow me in, please?"

He smiled warmly and said, "I just assist people to go through the correct doors to the existences that match their spiritual evolution. I send my blessings to you, Saho. Till we meet again! Goodbye!" And then, as though I had just asked a question that I was not allowed to ask, I was abruptly sent back to my physical body. (Interestingly, when you ask a question that you are not supposed to ask, one that exceed the bounds of knowledge permitted to you, while traveling energetically, you are either abruptly sent back to your physical body or you receive an answer in a soft murmur accompanied by a peculiar buzzing sound.)

Helpful Tips

Traveling before death

One's inherent ability to travel energetically tends to come to the fore when one is approaching death for three main reasons. First, it is only after one masters the physical senses and casts aside the physical body with the understanding of its function during this lifetime as simply a vehicle, that one becomes capable of escaping it. Second, the condition of the physical body upon death allows it. In most cases, when we are approaching the end of our journey here on earth, our physical body is terminally ill, irretrievably damaged, or both, and the energetic body escapes the physical body as an attempt to evade the sensations of pain. This ability of the mind to send the energetic body out of the physical is aided by the "fight-or-flight" response, which is conveniently ingrained in every person's mind and instincts. The third reason is that when we are approaching death, our mind is slowly altered from a strictly physical mind to a completely spiritual mind, thus making it easier for us to detach from the physical body.

The last two reasons form part of what I call the *physiospiritual automated system*. This is an automated system that is activated during one's transition from a physical existence into a spiritual one, altering the human physiological system and changing it to a solely spiritual one. This system can be triggered in various ways and by different factors, but is most commonly activated when one is approaching death.

The physiospiritual automated system can be best explained through the evolutionary process of adaptation. The human mind has successfully evolved to progress from a basic physiological function to a spiritual entity that possesses what we deem to be spiritual abilities at the time of death, all in order to adapt. As we near the end of our physical journey, our automated system is what prepares us for our adventures in the energetic realms. It is during this particular process of alteration that our spiritual abilities are heightened, explaining clearly why people who are approaching death often find themselves able to communicate with or see beings in different dimensional existences. Despite possessing no knowledge of this whatsoever, the nearly departed become natural experts in the adjustment of their energetic frequencies to freely communicate with and in the various dimensions.

Traveling to loved ones

The most difficult experience that we must all inevitably endure in the physical world is the loss of a loved one. When we lose a loved one, we believe that we have lost him forever, eternally, when in fact the concept of eternity is derived from an incorrect understanding of time. In actuality, time is a concept that we ourselves have created, and our perception of it is closely tied to our emotional state. As a clear and familiar example of this, you can look at your own perception of time and observe how it varies according to activities, events, and circumstances. When you're enjoying a vacation free from any distractions or worries, time appears to fly by quickly. Conversely, when you're in the middle of a grueling 12-hour shift, time tends to pass agonizingly slowly. This indicates that time itself is relative and elastic. Therefore, it's not surprising to find time moving unbearably slowly after losing someone you love and realizing that you have to go through the rest of your life without him or her.

In one of my earlier journeys I met a mother who had ended her own life by committing suicide. In this case, the elastic nature of time certainly played a part in her suicide. The mother's life in the physical world had been plagued by the thought of having to live the rest of her life without her beloved daughter, who had passed before her. Speaking hypothetically, if a higher being had visited this bereft mother before she killed herself, and informed her that she was going to die naturally very soon, her perception of time would have been drastically altered. Instead of committing suicide, perhaps she would have chosen to live out the remainder of her short life to its fullest. The reality is that most of us feel completely separated from our loved ones after they die, due to the barriers of both time and space; this perception becomes evident in the notion that we must wait for many years (perhaps a lifetime) before we can be reunited with them, at the time of our own death.

It is nearly impossible not to have heard of the Mayan calendar by now and its infamous end date of December 21, 2012. Many ignorant people have since dismissed those "ignorant, aboriginal Mayans" and how they supposedly failed to predict the end of the world. But the fact is that Mayans were not attempting to anticipate the end of the world with the ending of their calendar. Quite the contrary: The Mayans had long foreseen the rise of human consciousness that would halt our perception of time in the physical sphere. In actuality, the ending of the Mayan calendar is a brilliant metaphor that highlights both the intuition and great wisdom of the Mayan people. The Mayans were well aware that following December 21, 2012, most of humanity would leave behind their preconceived notion of the concept of time, thus lifting the barriers created within our minds that had previously hindered us from discovering the realities of other dimensional existences. The barrier of time simply creates an obstacle that prevents you from being able to see your deceased loved ones, and it alone imprisons you and blocks you from your true abilities.

The death of a loved one is only the death of his or her physical body. It marks the rebirth into an energetic existence. Though it is normal to feel immense loss and shed tears following the death of a loved one, you must realize that this person is merely residing in a higher spiritual plane of existence than your own, and that by raising your vibrational frequency and altering your perception of both time and space, traveling to him or her would become entirely possible. We are all multidimensional beings, free to roam and explore among the differing dimensional realms, and our existence does not solely lie within the third dimension, the physical.

In traveling to meet your dearly departed, the law of synchronicity always applies. Unless your plans are in perfect accordance with synchronicity, you will not be able to make your journey. The law of synchronicity is strongly connected to the concept of perfect harmony, and it is in force in all the energetic realms. This law is the very foundation upon which the concept of destiny rests—namely, that nothing can happen without there being perfect harmony in accordance with the law of synchronicity. Interestingly, the law of synchronicity can also be observed in the Fibonacci sequence, an observed mathematical equation that further illustrates how everything within this universe, whether it be an order of integers, a series of musical notes, or an arrangement of molecules, is imbued with and arises out of perfect harmony. Given this, if you are unable to travel to your loved ones in the beginning do not blame yourself, as it simply may not be the right time.

Some basic tips for visiting loved ones:

* **Know that if it is meant to be, it will be.** Much like the uniqueness of thumbprints, energetic coding differs from one individual to another. Because you are already familiar with the energetic coding of a close relative, the connection of your energetic body to his or her own will happen automatically. If you had a positive experience with this person on earth, then you will undoubtedly meet again!

* **Know that it takes two.** Visiting someone who has died is impossible unless both parties consent to the meeting. This is the case because the universe also observes the law of free will.

* **Use the power of intention and attraction.** As it is almost impossible for us to know the precise location of

our deceased loved ones, we must rely on the universal power of attraction when we intend to visit them. Through energetic intention, which is strengthened through the continuous disciplining of the mind, we can use the universal power of attraction. For example, when we are in complete control of our mind and thus not allowing negative thoughts to manifest, we are retaining the energetic power of creation that exists within our mind. This particular power will allow us to use both the power of intention and attraction to assist us in traveling out of body.

✳ **Respect the synchronicity of timing.** As I've already mentioned, everything in this universe occurs within the perfect synchronicity of timing. It is only when timing is in perfect accordance with synchronicity that an intended event will manifest. For example, sometimes you may be unable to reach a loved one simply due to the fact that there are certain trials or lessons that he or she must endure or learn. If this is the case, you ought to fully accept it rather than attempting to continuously contact this person, which will only result in his or her being constantly held back and distracted in order to answer your "calls." Now that you know this, don't be dismayed if you are unable to reach someone, as I was with my grandfather, for whatever is meant to be, will be!

3

Journeys to Familiar— and Unfamiliar—Places

It is human nature to yearn for what we can see but are unable to reach. We have successfully sent a few hopeful representatives to the moon, yet none of them was able to detect any form of life there. The contradicting fact that we have achieved advancements and innovations in technology and engineering, yet remained rigidly unable to open our minds to the vastly rich realities of our universe, remains a troubling thought, even if it comes as no surprise. In that sense our species is astonishingly similar to an insect that has climbed an apple tree and successfully reached an apple, but still finds itself unable to actually taste the fruit. The underlying question, of course, is why does the insect continue to roam aimlessly around the outside of the apple instead of piercing a hole in the fruit and tasting the sweet juice inside?

In order to attain full and complete knowledge of the energetic realms, you must tap into your consciousness and focus on linking it to whatever it is you wish to learn. As you gain mastery over your energetic body, you will find that this can be achieved by connecting your consciousness to the particular area of inquiry in which you wish to enrich your knowledge. Because the basic energetic reality of all beings, objects, and even planets enables this linkage to occur, everyone possesses the universal ability to energetically link themselves to any planet they wish to travel to.

Planet Mars

On the night of a full moon, my dear friend (and coauthor) Katy and I were sitting in the Jacuzzi chatting about astronomy. Katy is quite the astronomy enthusiast, so rather than obsessing over boys in our spare time, we spend most of our time together discussing the planets, stars, and galaxies as we stargaze late into the night. As Katy was immersed deeply in the topic of the possibility of extraterrestrial life, I suddenly blurted out in a serious tone, "Of course there is! There exist many different energetic realities, and each is categorized by its own distinct vibrational frequency. I travel between these dimensions, you know," at which point Katy's jaw instantly dropped. I noticed her begin to slowly ease away from me toward the opposite end of the Jacuzzi and heard her mumble something about having to get up early the next day. I let out a sigh, skillfully faked a laugh, and exclaimed reassuringly, "I was just joking!" She immediately seemed relieved by my white lie and was soon back to her usual self, responding to my false confession in her recognizable sarcastic manner, "If you travel between the dimensions, it means you've been to Mars and Jupiter then, Miss Jean Grey, and you've probably kissed Wolverine, too!"

What Katy didn't realize then was that I was telling the truth: I have indeed been to Mars. I traveled there on the night of my 19th birthday, determined to make my birthday night an especially memorable one. You could say I succeeded, but it sure didn't come easily. I had previously attempted to travel to planet Mars on several different occasions, but to no avail; each time I was abruptly and rather discouragingly sent right back to my physical body. It was only later on that I learned that the universe functions in mysterious ways, and when a journey is not fated and hence blocked, it only means that it is not the right time. Nevertheless, the night of my 19th birthday turned out to be just what I had deeply longed for, and after performing the usual protocols (see the Appendix for a short reference list of these) I was able to successfully travel to planet Mars.

To my great surprise, the planet's energetic existence closely resembled the photographs we often see of it. I found myself in the center of a vast, empty space, surrounded entirely by a peculiar maroon-colored surface that appeared to be made out of packed grains of sand. The atmosphere was rather murky and dim, and all I could see was a calm, dark ocean, which actually looked more like an unusually large lake. With my curiosity taking a toll on me, I began to wander aimlessly around the empty space, until I noticed a palm-sized cube-shaped device maneuvering itself over the planet's sandy surface in a peculiar manner. Its movement appeared rather awkward and unpredictable in nature, leading me to cautiously suspect that it was most likely a tracking device. As I contemplated whether or not I should approach the device, I felt a forceful blow across my back, but before I could turn my head to find where it had come from, I was tightly blindfolded and brought on to what seemed to be the back of an animal.

Once the blindfold was taken off of me, I saw a dark red female being standing right in front of me. She peered at me through her peculiarly large eyes, as though expecting an explanation for my presence. But I remained mute in front of the sight of her towering height and her arms that extended down past her knees. Her clothes resembled those of an aboriginal warrior and were made from a dark red metal that almost perfectly matched the color of her skin. I realized almost immediately that we were inside some kind of hut, and that everything surrounding us was dark red in color, too, including a stack of chairs, their seats the shape of hexagons, that was placed next to a large door. Eager to break the awkward silence, I introduced myself telepathically to the being: "My name is Khartika. I come from planet earth. I meant no harm. I'm a traveler." She did not reply, and instead began to frantically walk in circles before finally deciding to approach me and embracing me tightly. It has always been rather difficult for me to put into words exactly how I felt in that particular moment. Her embrace seemed to have awakened a pleasant surge of vibrant energy within my energetic being, and I felt a deeply overwhelming sensation of love and purity flow freely through me.

As I remained immersed in this beautiful feeling, she quickly moved toward the door and brought back two hexagonally shaped chairs, instructing me to sit down. I could not help but wonder why she had decided to hug me, as I had never experienced anything like that in any of my previous travels. When I asked her about this, she replied in a slightly nasal tone, "Here in Mars, my kind decide whether one is to be trusted or not through forming an energetic connection with them. Every being possesses its own unique energetic coding, and it is this energetic coding that emits the sound that your kind probably refer to as music. This music is unheard of by most beings, but to my kind it is heard through the gesture of energetic alignment, or what you call a hug. Now I know that you are to be trusted. My name is Danau. Why are you here? If I had not come and taken you with me, you would have been attacked. Do you have no knowledge of the Qeros? They detect any energetic change in our planet. If any unusual energetic fluctuations are detected, they will automatically emit an electromagnetic force that could truly be detrimental to you!"

She suddenly jumped up from her seat and pulled on a small lever that was dangling in the middle of the room, and the dim lighting in the hut was instantly switched off, leaving us in complete darkness. "It's safer this way," she explained, quickly picking up on my inquisitiveness. "This way, they won't suspect that you're here. The Qeros sent the patrols for you because they probably detected an energetic fluctuation. Darkness has overtaken our planet. There are two factions existing in Mars now; one is the ruling faction, who are puppets of darkness, while those of more spiritual nature, my kind, remain in constant hiding. We are outnumbered by far."

"Then our meeting is most certainly fated, Danau," I exclaimed. "I came here because I have always been inquisitive about your planet. As you know, your planet exists in the physical reality, too. But most importantly, I have been traveling to different realms to alert them of the force of darkness that has overtaken planet earth. These humans are living in a state of utter illusion; they do not know of the realities and, most importantly, they are not aware of what's happening energetically to their very own kind. They are satisfied living in this physical bliss without seeking the truth. They are in a trance, Danau!"

She reached across and took my right hand, placing it gently on her forehead as she kindly said, "I sympathize with you. I read all of your thoughts and chatters. I sympathize with you. May a candle be lit on your planet earth." She then proceeded to lift her right hand and place it on my forehead, as she continued explaining, "Khartika, our planet, along with the planets that you know of as Jupiter and Venus, originally existed in a higher dimension than planet earth. But we were manipulated by the darker forces to exist in the same dimension as planet earth and the physical planet earth, for their own uses."

"So, you currently exist in the same dimension as the energetic earth? And your planet's physical existence occupies the same dimension as planet earth?" I asked, perplexed at the realization her words had brought me to.

"Our planet, along with the physical planets that humans know of, exists in the physical—in the same dimension as planet earth. But a small number of planets along with ours also exist in other dimensions as well. That is the reason you are able to interdimensionally travel here." She paused for a few long moments, as if turning a thought over in her mind, and then continued hurriedly, "You must go now! It is not safe. May a candle be lit!" she said in a worried tone. With her alarming words, I felt a wave of tingling energy quickly migrate through my forehead, automatically forcing my eyelids to close, and when I managed to finally force my eyes open once again, I found myself back in my physical body.

The Moon

I must admit that my visit to the moon is one that still perplexes me to this day. My friend Mary, who had received the rather loud wakeup call from reality while we were on vacation in Bali, had recently grown obsessed with the moon. She had just returned from her trip to Hong Kong and arrived with a suitcase full of "moon dust" souvenirs in tow. These souvenirs are pieces of interesting-looking gravel that have been placed in small glass containers, which, according to Mary and the vendors who fervently sell them, are dust fragments from the moon that will bring fortune to their owners. As Mary handed both Jane and me a moon souvenir, with an undeniably proud smile on her face, I noticed that she was basically covered in everything moon: She wore a dark t-shirt with a bright image of a full moon plastered in its center, a silver necklace with a dangling half-moon pendant, and even a ring that looked like a full moon and made entirely out of glass!

"You look so vibrant after your trip, Mary!" I complimented her.

She giggled with the same beaming smile and replied excitedly, "Yes, you have no idea! I uncovered amazing revelations about myself while I was in Hong Kong. I was shopping in the old market, when an old man asked me if I wanted a reading, and I thought, *Why not?* So he gave me one and he told me that I had a past life on the moon, and that my soul mate currently resides there. He also told me that the moon itself is actually my lucky charm, so I have the most luck whenever there is a full moon! Remember how I got an A on that political science final exam I didn't even study for? No wonder! I remember looking up at the sky that same night, and there was a freaking full moon!" As I tried to form an appropriate reply, Mary quickly added, "Oh, I forgot to say, the old man also told me that I was a lama of the high mountain in my past life, too, and—"

"You were a *llama*?" Jane interrupted, her forehead crinkled in bewilderment. "As in the animal?"

Mary, who seemed deeply offended by Jane's innocent remark, replied snappishly in a cold tone, "No! A lama, as in a monk of the higher order—*obviously*."

I remained silent, trying hard to contain my laughter while Mary resumed her passionate speech about her past life on the moon. All I could think of at that moment was how shocked Mary would be if I told her the truth about the moon.

As much as we desire everything to be perfect, imperfections will always exist. This is a fundamental concept that applies to everything, including multidimensional travel. When I first started to entertain the thought of traveling to the moon, my stubborn curiosity, which I've already mentioned a few times, got the better of my intuition and misgivings, which told me that I shouldn't go.

From the very beginning of my journey, I was faced with obstacles that I could not comprehend. Even though I had carefully directed my thoughts to travel to the moon, my energetic body remained there in my own room. After some effort, and through sheer willpower, I managed to guide my energetic body to automatically align itself with the same vibrational rate as the moon. I was horrified and left speechless by what I saw when I arrived. An alarming number of what looked like spaceships were orbiting the moon, their irregular and rapid movements shattering the field of my vision. On the surface itself, I could make out a group beings that appeared to be desperately trapped in some kind of invisible force-field. They were all speaking among themselves and engaging in a variety of odd tasks. I spotted one being, which strongly resembled the energetic body of a man, anxiously roaming back and forth while gesturing with his hands, as though he were speaking to someone or something that could not be seen. His gestures grew even more peculiar as he began to make swooping movements with his hands, as though he were digging.

Even though I stood right in front of him, he could not see me. It was as if there was an incredibly thick and tall wall between us. Waving my hands over my head, I desperately tried to catch his attention, until I suddenly received a telepathic message: "Over here," a soft female voice echoed through my mind, attempting to divert my attention. A few feet opposite from where I stood, I noticed a woman sitting cross-legged on the ground. She was dressed in what appeared to be a military-type uniform in tones of green and brown that blended in with each other, and her dark blonde hair was bundled in a messy bun on top of her head. Still extremely curious about the invisible barrier, I extended my arm outward to try and touch it when I heard her now-familiar voice telepathically ring through my mind: "You do not want to touch it!" I immediately jerked my hand back and jumped backward. "Everyone here, aside from me, is trapped in their own time and their own illusion," she began to explain. "None of them is entirely conscious of their true whereabouts. I just managed to break free of what I've come to call the robotic dream sequence. I don't know how long it will last before they find out that I've broken through their wavelength emission." I peered at her uniform, and the woman continued telepathically, intersecting the course of my thoughts, "Yes, my physical body may be dead by now. But it may still be alive; I do not know. Many people that experiment with matters of time and space tend to find themselves on the moon. Even I can't explain it. See that gentleman with the black hair? We ended up here together." She pointed at the man, but I paid no attention. I *thought* I understood all the fundamentals of multidimensional travel, yet I could not wrap my head around what she had just explained to me.

"What is the moon? Who resides up here?" I finally decided to ask her in my confusion. Suddenly I saw her desperately grasping her chest in obvious pain. She let out a loud cry of despair as her body was lifted off the ground by some unknown force and flung toward the invisible barrier. She fell to the ground with a heavy, agonizing thump and seemed to remain unconscious for a few seconds before waking up and beginning to move around in a strange manner. Like the oddly mannered man who had caught my attention earlier, she seemed to be talking to herself as she went through the pantomime motions of preparing a meal.

Shocked and confused, I desperately tried to get her attention, but soon realized that there was no use. She was now stuck in the same trance state as everyone else. I thought about calling on the higher beings for assistance, but before I could even manage to try, I felt myself being vacuumed upward by a strong, unseen force. As I was pulled higher and higher off the surface of the moon, I began to feel excruciating pain, like that from a hefty electric shock, course through my body. It felt as though I were being sliced open by a knife and zapped with electricity all at the same time. In panic and terror I called for assistance from the higher beings several times, but my cries went unheard, and the unbearable pain I felt within my being escalated exponentially. I was left with no choice but to do what any desperate multidimensional traveler from earth would do: I began to focus my mind on contacting my physical brain so that it would alert me to wake up. "Wake up! Wake up! Open your eyes!" I shouted telepathically to myself. I felt my eyelids begin to lift themselves in infinitesimal fractions, as my right arm gravitated to my right eyelid to force it to open. But I could not. My consciousness had only managed to transfer itself to my physical brain for a few seconds before returning to the same energetic situation. "You have to wake up! Wake up! Wake up! Wake up!" I anxiously screamed to myself, and as soon as my consciousness finally transferred back into my physical body, I threw myself to the ground from the edge of my bed, thinking that the shock of the fall would force my energetic body (or consciousness) to return to its physical existence. I forcefully pinched myself and even slapped my face several times, until, getting desperate, I finally grabbed a bottle of water and poured its contents over my face to wake myself up completely. I found myself drenched in not only water but an alarming sensation of fear, so I stumbled for my phone and quickly dialed my mother.

"Hello, dear? Isn't it 5:30 a.m. in L.A.? Have you been staying up late again? I told you, it is not good for your health! Staying up late will lead to kidney failure, which will lead to you being sick, which will lead to high medical bills, which will consequently lead you to dying early..." As she continued to lecture me, I wondered if she was planning on taking a breath anytime soon.

Once I had an opening, I replied to her reassuringly, "Yes, I'm fine, Mom. I just wanted to see how Grandma is."

~~~

Because we live in a mysterious world that we are only beginning to truly understand, you will find yourself faced with unexplainable situations that, armed with your previous knowledge alone, you are simply incapable of handling. With this in mind, it is crucial that you prepare yourself for *any* eventuality during your multidimensional journeys. When you travel energetically, your energetic body possesses the greater part of your consciousness, while your brain retains only a mere fraction of it. During unpleasant or dangerous journeys, always keep in mind that you can transfer the larger portion of your consciousness, the one that resides in your energetic body, back into your physical brain so that it can alert it, awaken it, and (hopefully sooner rather than later) pull your energetic body back into the physical. Wherever in the universe your travels may take you, you will always have this invisible "walkie-talkie" available as a medium of communication. As long as you focus on remaining in communication with your physical mind, your "command center," during your multidimensional travels, your brain will act as an incredibly reliable conduit for receiving these messages. And here is where the beauty of being an energetic traveler truly lies: Regardless of the situation, you will always have this safety net available to you.

I'm sure you are curious about what happened to the people or beings residing on the moon. My best guess is that their physical bodies lost their immediate connection to their energetic bodies when their energetic bodies traveled or were transferred to the moon. Consequently, their safety net had essentially been destroyed. Such is often the case for most people following their physical death: They remain stuck in a state of limbo, neither here nor there, and because of this they have yet to realize that their physical body has ceased to exist.

Consciousness

Consciousness can be thought of as the invisible drive behind one's spiritual life-force, and it can only be elevated as one becomes more spiritually attuned with the energetic, or unseen, world. In most people, the majority of their consciousness resides within their mind; but this part of their consciousness represents only a small fraction of the whole. When most people are asleep, their consciousness remains within their physicality, untouched and unused. However, as someone masters and attains the true understanding of multidimensional travel, his consciousness will become elevated, thus allowing a fraction of this consciousness to be transferred into his energetic mind, and thereby making him much more aware of his energetic adventures. He will be just as awake while traveling through the energetic worlds as you are right now as you read this book. And he will be able to remember much more!

If you are one of the few people who are blessed with the ability to remember most of your dreams, you have already received the larger slice of the pie, for your consciousness is more elevated than that of the majority of people. Your goal is to elevate this fraction of consciousness high enough so that you are able to travel in full alertness while still maintaining communication with your physicality as a means of safety.

Helpful Tips

Categories of multidimensional traveling

* **Full awareness.** The first category of multidimensional travel is the most typical method used for traveling in the energetic realms, and occurs when an individual is in an elevated state of full awareness. In this category, the

traveler is able to fully transfer her consciousness from within the essence of her physical mind, where it usually resides, to her energetic being. You will know you've traveled in full awareness when you are able to recall the whole journey, from when you first detach from your physical body to the moment you return to the physical.

* **Partial awareness.** This category of traveling occurs while dreaming. When people dream about their deceased loved ones, more often than not, they are traveling energetically in partial awareness, though most of them remain unaware of this. It is in the dream realm that most people communicate with their deceased loved ones. What is actually occurring here is that they are unconsciously using their power of creation to rearrange the energetic particles of the human energetic dream realm. If it's helpful to you, think of it as a limbo state.

Envision an untamed, wild animal: This is your power of creation. You can either tame the animal and teach it to be your protector, or you can simply let it run wild. Choosing the latter option is basically choosing to manifest your subconscious nightmares and forcing your energetic body to put up with them. Because the human dream realm lies in the energetic existence—specifically, within the lower sector of the fourth-dimensional realm—your deceased loved ones (who have already passed through various realms) already possess the ability to visit you in your dream state. Unfortunately, because you are dreaming, the forces of your own creation disrupt your communications with them and cause you to perceive the whole thing as a dream rather than reality. My friend Jane recently provided me with a good example of this when she described a dream she had in which she conversed with her dead mother. She insisted that it was "only a dream," as she recalled her mother reminding her to take good care of herself and remember that there is life after death. As Jane's experience reflects, it is crucial that you understand that this power of creation can either be enhanced (and used to your own advantage to help you understand the nature of the universe) or ignored, in which case it can quite easily become your greatest enemy.

✳ **Redirect.** This third category of multidimensional travel is a new form of energetic travel I discovered that has evolved over time. The conscious travelers of the past were able to freely and easily use the first category of multidimensional travel, but because we live in different times, times that are as a whole hostile to the idea of the energetic realms, the first method may not work for you (that is, unless you desire to travel to the lower-dimensional realms, which I doubt).

The fact is that our dormant ability of multidimensional traveling is being continuously suppressed and hindered by dark forces. Have you every wondered why most people have more bad dreams than pleasant ones, or why some mornings you wake up feeling extremely drained and exhausted rather than energized, even after a long, good night's sleep? The underlying reality behind these puzzles lies in the fact that even if you follow all of the appropriate protocols and you discipline both your mind and your heart, you may still find yourself unable to journey as you wish and instead experience continuous subconscious dreams. Sadly, this has become the norm, but the fault is not entirely our own; not only is the ability to travel energetically being suppressed, but our brains are also being externally stimulated for the purpose of automatically inducing these usually negative dreams on a constant basis.

In order to travel via the redirect method, you must learn to redirect your consciousness, which resides in the physical mind, to the energetic essence of your being during the so-called dream state. The key to doing so lies in the ability to successfully alert your consciousness of your whereabouts, which in turn will stop the automated creation of your dream. Once you are able to do this, you will be able to break free of the human energetic dream realm. This method is not easy to understand, let alone master; however, with great dedication and focus, you, too, can learn to redirect your mind in order to eventually venture out into this universe of limitless possibilities. As a higher being once said to me, to understand is to know, to know is to feel, and to feel is to experience, but experiencing is not enough without constant dedication and connection to one's true essence.

Protocols for redirect journeys

1. **Align your thoughts with your intentions.** When you align your thoughts with your intentions, in actuality you are coordinating your mind with your heart. Therefore, by following this protocol, you are activating your ability to differentiate the subconscious dream realm from your own physical reality.

2. **Learn to distinguish between dreams and reality.** In order to be able to alert your mind to the true nature of your experiences, it is crucial that you learn to have discernment regarding your experiences in the dream realm. Although it is often extremely difficult to discern

this ersatz reality from the experiences you have in the physical world, you must not give up. It is only after you have successfully mastered identifying the human dream realm that your mind will be able to redirect its consciousness to your energetic essence.

3. **Focus.** Once you have realized that your dream is indeed only a dream, focus your thoughts on where you want to go. Concentrate on intense feelings of love and light, as you direct your thoughts toward the particular planet or realm you wish to visit.

Exercise

First, envision a balloon of the color of your choice floating through the air, and concentrate on it with your complete focus and intention.

As you continue to visualize the balloon drifting slowly into an expanse of sky, simultaneously begin to envision yourself sensing your surroundings. Smell the distinct salty scent of the ocean, and hear the crashing sound of the waves and the rhythmic chirping of the birds. Feel your entire presence there by using all of your senses fully.

Now, take a deep breath and tell yourself that everything you have just perceived is *not* real. Envision yourself escaping this scenario by lifting yourself off the ground and directing yourself back to your body.

When engaging in this exercise, you are using the most important component of the redirect method of multidimensional travel—namely, your mind's ability to distinguish between your own subconsciously self-induced dream state and the realities of all beings. Exercises such as this teach your mind how to effectively differentiate between the true energetic essences of realities while in the state of sleep. As you continue practicing, use your own creativity and create different scenarios, but make sure they are realistic. An important component of this exercise relies on your ability to create scenarios and inner narratives that are actually possible, as opposed to those that are obviously fantasy. For example, if you envision yourself roaming around the chocolate factory with Charlie, or picture yourself as Professor X from *X-Men*, the purpose of the exercise will be entirely defeated, as your mind will naturally be able to detect the irregularities and oddities of such scenarios and thus understand that it is simply an exercise for tricking the mind. The key is to construct scenarios that make you feel lost in the moment and that drawn on all of your senses. My closest friends have found that the most useful and successful scenarios for tricking the mind are those that incorporate real people from their daily lives, such as their annoying mother screaming at them during breakfast or their crush taking them on a date and kissing them under a tree. Whichever scenario you choose, it must have the element of probability.

4

Rhythms of Life

"Oh no! What *is* it about today?" Mary let it out in an exasperated sigh as we strolled down Santa Monica's jam-packed Third Street Promenade. "Why is it that whenever I put my sunglasses on, it always turns cloudy?" I stared at her with the blankest expression I could manage to pull off. Mary always made me laugh, even when she wasn't trying to. "What? You don't believe me? Here, watch!" she demanded in an irritated tone that threatened she was not one to be mocked. She quickly slipped her trendy bug-eyed sunglasses back on, but soon enough was distracted by the sight of H&M, her favorite clothing store. There was a sale going on, which meant we'd have to go in.

Fifteen minutes later we were back outside, with four of Mary's new shopping bags in tow, only this time the golden California sun was nowhere to be seen. It was pouring buckets, and just about everyone was seeking refuge in the local coffee shops and bookstores. As I looked up toward the sky and felt the cool raindrops make contact with my warm skin, Mary smiled proudly at her victory of being right. I let out a joyful laugh of relief; it had not rained in Los Angeles for a while. But more importantly, the rain always meant that I would be able to travel undisturbed through space and time for as long as my journey was fated to last.

Vibrations

Rhythmic vibrations consisting of energetic frequencies exist everywhere around us. In fact, your very own energetic body releases a sound that is personal and unique to you, and your every action elicits movements that release these vibrations. Although all of us emit these rhythmic vibrations, each and every one of us does so within a specific range. Because we cannot hear these vibrations with our ears, we must instead intuit them by utilizing our energetic essence; in this way we can detect the undetectable.

Have you ever wondered why listening to certain songs makes you feel extraordinarily happy and even electrified, while others seem to draw your being down with negative emotions? The reason behind this phenomenon does not simply rest on the fact that each of us possesses our own unique taste in music. Music that you find yourself strongly attracted to and constantly replaying on your iPod actually sets off an energetic sequencing in your energetic essence, which then correlates with your overall vibrational frequency. It's not just the lyrics that create the mood, but, more significantly, the rhythmic vibrations that cause energetic particles to connect with your energetic essence. Many people claim that you are what you eat, but this only pertains to your physical body and not your mind. In actuality, you are very much what you listen to, and you will surely come to understand this through your many multidimensional adventures.

Music itself is an energetic resonance that connects your essence with its surroundings for the purpose of forming a rhythmic vibration. The adepts of the past were very much aware of this; thus, they did not play music for the sake of pleasure and leisure only. Instead, they used music to elevate their overall vibrational frequency and heighten their levels of consciousness, which enabled them to travel multidimensionally through the governing of their energetic bodies.

After spending an hour or two in a busy or hectic environment—a bustling college library, for example—we often find ourselves feeling inexplicably drained and exhausted without knowing why. The first explanation for this rests on the fact that people who vibrate at a much denser (more negative) vibrational frequency than you unconsciously sap your positive energy. In this sense, energy in the energetic realms follows the laws of thermodynamics. The second reason behind this common problem is the general disharmony that often exists in a crowded room. Because spirituality is rarely emphasized in the modern world, a room full of people generating extremely diverse frequencies can contribute to a highly jarring, discordant overall frequency. Just imagine the disharmony that results when all the lower piano keys are played at the same time. In such situations, your energetic body translates this disharmony to your heart, which then creates an intuition that informs your mind of the unpleasant and draining feeling of the room you are in.

Sensing and Intuition

Once you are able to discipline your mind and control every thought that enters and exits, and maintain only thoughts of positivity, your overall vibrational frequency will rise drastically. And once your overall vibrational frequency has been elevated, your heart will begin to assist in maintaining this elevated rhythm. Consequently, the connection between your mind and heart will be solidified, and eventually, through its linkage with your mind, your heart will be able to detect energetic frequencies in others that you would otherwise be oblivious to while in your physical body. In short, you will become exponentially more intuitive.

Music as a Tool

Music is the primary universal tool used for communication in the higher realms, where it exists as a united rhythm composed of elevated rhythmic vibrational frequencies. The higher dimensional realms work together as one, and it is precisely this unity of Togetherness that forms a unity of rhythm through the interconnections of various, unique, positive vibrational frequencies. Rhythm is an energetic connection that unites and interconnects all beings residing in a particular dimensional realm, and it is this joint rhythm that enables them to assist each other in the elevation of their overall vibrational frequency, and that fosters and maintains the high states of existence in the higher-dimensional realms. During your travels, your energetic body will be able to detect the music—the combined rhythmic energetic vibrational frequencies— of all the beings of the particular dimensional realm that you're visiting. When you travel to the higher realms, for example, you will be able to detect a beautiful melody through your own connectedness with that planet's overall joint rhythm.

As you are falling asleep and changing into your energetic body, at times you may hear snippets of different types of music in the background, for only a few seconds or possibly longer. This music should not be ignored, as it serves as a useful and accurate signal for detecting which dimensional existence your energetic body is slipping into. Put in more simplistic terms, it can help prepare you by telling you whether you are going someplace good or someplace bad, which, as you have already learned, is very much dictated by your own state of vibrational frequency in that moment.

Dance

Dancers often stress that in order to learn a new dance you cannot simply memorize the steps; you must also feel or inhabit the rhythm. The ritualistic dances of aboriginal or more isolated cultures are not simply performed for the sake of fun, as most American dances are today. In fact, the dances depicted in many works of art recovered from important archeological sites illustrate the rhythmic vibrations and interactions that exist among all people. The reason these ancient civilizations left such clues for the future generations was because they wanted to portray the energetic vibrational world in a way that would encourage the observer to sense and even feel these shared rhythmic vibrations in the physical body. When they danced together in groups, it was a way for them to communicate the rhythm and flow of energetic unity to one other. This is why dance was so often depicted in ancient works of art.

With this in mind, it is not surprising that many people have used and continue to use group dance as a form of therapy to mitigate the stress of everyday living. Many such people have also reported feeling increasingly more in tune with themselves, as well as their surroundings, and in a better mood in general after participating in dance as part of a group. The reason for this is that dancers in a group form connections with one another as well as with the group itself, and this connectedness is reflected in a temporarily elevated level of intense energetic frequencies. It is important to note that in order for dance to have such an effect, individuals must dance in tune with the right type of rhythmic vibrations of energetic frequencies—in other words, the right type of music. When the correct type of music is used during a communal dance, the overall vibrational frequencies of the dancers are elevated as they create a joint form of connection with one another's energetic bodies.

Whenever you engage in any form of dance, whether alone or in a group, you are communing with the energetic world. It is no coincidence that even our brain is electrical in nature and responds naturally to wavelengths, frequencies, and electricity. Of course, not all dance is created equal; when you dance, it is best that you do so to music that genuinely makes you feel good in your heart and your mind, instead of defaulting to the popular music of today, which is actually misleading people, without their knowing it, through strategically deceptive lyrics.

Musical Adepts

As you know, ancient civilizations in general were remarkably knowledgeable about the true energetic and rhythmic existence of our universe. Thus, these adepts of old also possessed a true understanding of the concept of music. After acquiring knowledge from their travels to various dimensional realities, they quickly discovered that they could create musical instruments that would assist them in elevating their vibrational frequencies.

Although they were well aware that the best way to travel was through the elevation of their vibrational frequencies, they also lived (as we do) in the "real" physical world where suffering is ever-present. Thus, they knew it would be impossible for them to maintain a state of high frequency at all times. So they created musical instruments that would enable them to mimic the exact *mélanges* of sounds they heard during their travels to the higher-dimensional realms. From there on out, upon hearing the music in their subsequent travels to planets in the high dimensions, they would instantly feel a connection between their own energetic essence and that of the planetary existence. They knew very well that music was a tool that would greatly assist them in maintaining a high state of elevated frequency wherever their multidimensional journeys would take them.

A Boy and His Dolphin

On one particularly misty spring night, I managed to travel to a planet entirely composed of water—my perfect escape. All around me I saw complicated mazes of thin, interconnected wooden decks scattered over the glistening, calm aqua water. There was not a single wave in sight, and the only noise came from the light slapping sound of my feet against the damp wood. As I hopped from one deck to another, overwhelmed by a feeling of extreme joy, I suddenly lost balance and nearly skidded off into the water. Struggling to find my equilibrium, I was quickly distracted by a chubby man sitting at the edge of the deck with his legs dangling in the water. He had light brown hair, and was wearing shorts and a plain t-shirt, nothing too different from what you would find on planet earth. Leaning back on his elbows and with his eyes peacefully closed, he did not acknowledge my presence until I approached him and politely asked where I was. "The name of this planet is the Third Planet," he answered telepathically in a peculiar accent, stretching the vowel of each word as if his life depended on it. "It is here that animals and beings such as us live together in constant harmony. It is not like on earth, where you are from. Here, we live as equals with the animals as they roam about." He straightened his back and opened his eyes, staring into the clear water ahead of him. I could see the panorama of the deep blue ocean reflecting in his sparkling eyes, but before I could compliment the water's great beauty, he started to tell me a remarkable story. "Long ago during ancient times, there lived a boy in a certain region of earth. It is now known as Greece. He grew up a sad orphan, his life overwhelmed with countless misfortunes and obstacles. When he turned 8 years old, in what seemed to be yet another affliction, he slipped off the side of a cliff and plummeted into the depths of the ocean. In that moment, the boy was convinced that death was unavoidably near for him and that, indeed, his life was over. But instead, in the midst of being half unconscious, lying on the misty ocean floor surrounded by the deep-sea dwellers, he felt himself suddenly being placed on top of a large fish. Darting through the expanse of the blue ocean, the fish hurried to the surface, and soon enough, the boy managed to take a breath of air. He found himself on the back of a dolphin. Since that very incident, it was as though he had been transformed into an entirely new person. He felt more positive, more alive, and happier than ever before. But perhaps most impressive of all, he instantly began to emit a remarkably high vibrational frequency, and he could not explain why."

As the man paused, he looked out into the endless expanse of the blue horizon. I moved a couple feet closer to him and sat on the edge, gently dipping my feet into the warm water. I thought about the boy and the dolphin that had saved him; having just returned from a trip to Greece, I was quickly reminded of something: "Did you know that there's a very well-known image, originating from ancient Crete, of a boy riding a bottle-nosed dolphin and playing the flute?" I asked him in excitement. "Do you know what the image truly means?"

Turning toward me, a warm smile came across his face, and after pausing for a short moment, he lifted his forefinger and placed it between his thick eyebrows. "Contrary to popular belief on your earth, the odd and unusual-looking noses of the dolphins do not serve as any particular symbol. The image that you speak of depicts the real dolphins of the past on earth. It illustrates the strong connection, formed through the power of the higher-energetic frequencies, that exists between the boy and the dolphin. Dolphins are unique in that they are able to identify the overall vibrational frequencies and energetic essences of other beings. The boy was able to achieve a higher vibrational frequency through the presence of sound." Upon hearing his words, I immediately thought of the boy communicating with the dolphins through the flute and wondered if I, too, should try doing that. Interrupting my thoughts, the man explained, "It is incorrect to assume through this image that the dolphin was able to identify and connect with the boy through the sound emitted by the flute the boy is pictured playing. Dolphins are indeed attracted to par-ticular sounds and are most certainly able to identify specific sounds of higher nature or higher-vibrational frequencies, but their ability to do so is achieved through the usage of vibrational frequencies that they them-selves emit. These vibrational frequencies go hand in hand with water to achieve an energetic unity."

He paused briefly to take in a deep breath of the ocean air, before continuing slowly, "The boy...the boy in the image...he is playing the tune of the energetic bond that exists between him and the dolphin. This energetic bond can only be heard energetically, not physically. When you travel to various high-dimensional existences, it is likely that you will form friendships with the higher beings that you meet on your ad-ventures. You must be careful not to confuse these friendships with the superficial friendships that are common on earth, those that are usu-ally achieved through gossip or trivial connections. The friendships you form on your multidimensional journeys are instead achieved through similar energetic sequencings. They are bonds consisting of energetic alignments that go far beyond lifetimes and dimensional realities. When such a sacred bond is formed between two beings, the conjoining of their energetic frequencies creates a vibrantly beautiful symphony that is heard only in the energetic existence. Such a bond, in human terms, causes a frequency that is especially 'loud' when the energetic essences of each being are next to each other."

The man concluded his story, bringing his hand back down behind him, and turned to me with an expectant look in his deep-set eyes, as though waiting for my input. In my mind, I was captivated by the image of the kind dolphin and the boy. The beauty of their energetic relationship engulfed my every thought and I implored myself to remember the beautiful story. With a grin on my face, I promised him, "I'm going to write this in a book, but I don't know if I'll be able to remember all of it!"

He inched toward me and, placing his palm gently on my forehead, said, "Worry not, my friend, for the universe will make it so. Come; let me show you something. Follow me without fear." He instantly jumped up on his feet and dove into the ocean in a perfect motion, leaving a just a flick of splashing water behind him. Without any hesitation, I fearlessly followed him into the water, for I had always wondered how it would feel to be in the ocean while in the energetic body. There was no particular sensation that I could point to; immersed in the water, I felt absolutely no temperature—no cold and no heat—but an indescribable feeling of immense positivity soon began to spread within my chest.

As I followed him toward the bottom of the ocean, trying to copy his graceful strokes and movements, I was shocked to see a wide, vacuum-like space that was separate from the rest of the ocean and completely devoid of water. He telepathically communicated to me, "This vacuum exists on earth, too." We swam further toward it when, suddenly, I was able to see what looked like virtually every species of animal—everything from frogs to dogs—roaming around within it. I came to a stop, astounded by the magnificent site in front of me. The man stopped as soon as I did and let out bubbles of laughter: "You are thinking about how they can breathe here? I thought that you're a multidimensional traveler! How can you still think of oxygen and all those human physical needs? I thought you had surpassed those thoughts. As I told you, we live in absolute peace here. There are many beings of love and light existing within the vacuum of earth's ocean, too." I peered at the animals in absolute awe one last time, before waking up and writing this chapter.

Everything made sense to me after I returned from this aquatic planet. I had long heard of numerous ancient musical instruments being uncovered from the ruins of past civilizations, many of them resembling modern flutes. Various archaeologists have attempted to test these uncovered instruments through the years but, much to their dismay, have been unable to create any sounds with them. Of course, this begged the question of why these cultures created instruments that didn't make any sounds. Some thought that perhaps these instruments were missing critical pieces. But in the journey I just described, I learned that this deduction completely misses the simple reality of these ancient flutes: They were in fact ancient symbols of the strong connections people had with higher-dimensional beings in the energetic realm. The wise man's story reminded me of the friendship bracelets that are popular today, illuminating that, just like these bracelets, these sacred instruments were used by ancient civilizations to remind them of the unique and sacred bonds that existed between themselves and the higher-dimensional beings.

Our Bond With Water

The strong bond that exists between humanity and water has persisted throughout time; however, the true potential of water as a rhythmic, energetic force is, unfortunately, yet to be recognized by the present generation due to a general lack of spiritual knowledge.

In order to understand the true significance of water, it is useful to observe your own relationship to it. Whenever you are in need of emotional support, do you find yourself naturally attracted to or seeking solace in a locale that boasts an abundance of water, such as the beach, a lake, or a river? Whenever you wish to ignite or nurture your creativity, seek solutions to your problems, or even receive inspiration, do you tend to gravitate toward places that are close to water to do so? What about when you are with your mate? Do you enjoy spending romantic time near the water? If any of these ring true for you, then you must realize that it is your energetic, intuitive nature that is speaking to you and offering you an intimate understanding of the power of water.

Within its essence water possesses the true components of love and light, which enable it to function as a significant uniting factor for energetic frequencies. Water is very much connected to the power and unity of Togetherness: Alone, it fosters the successful harmony and elevation of energetic frequencies; as a unitary whole, it is a seamless combination of discrete energetic frequencies.

As dolphins are true beings of immense love and light, their essences vibrate at an especially high and constant energetic frequency. For this very reason, their existence on our planet, in a much lower-dimensional existence than their place of origin, can only be supported in water. As previously explained, in order for a being to be able to use and sharpen its ability to sense through energetic intuition, the being must possess an elevated level of vibrational frequencies as well as the force of love. A dolphin's ability to both sense and feel is significantly intensified in water; in fact, as beings that possess intimate knowledge of the energetic realities, their primary means of communicating with one another occurs through sensing and feeling. Dolphins further serve as enhancers of the elevated rhythm of energy that already exists in water, effectively helping enhance the vibrational frequency of our planet's energetic sphere and remaining entirely unnoticed while doing so.

Minoans and healing

The only true form of healing that exists is that which occurs via the use of water. By this, I do not mean that you should immediately begin splashing water on your newly divorced and depressed sister's face in an attempt to heal her! Instead, I intend to emphasize that the one true form of healing is one that embraces and appropriates water's energetic nature.

Of all the civilizations that have existed here on planet earth, the one that I have always regarded as the most spiritually attuned is the civilization of the Minoans. The Minoans lived in absolute harmony with energy. The high value that they evidently placed on agriculture serves as a clear indicator of how they were able to achieve such elevated spiritual evolution, while other civilizations failed terribly at this. In order to live in complete harmony with energy, as the Minoans did, one must truly understand the essence of water and, in doing so, cultivate a sincere respect and appreciation for it.

The Minoans used water in the energetic sense, for the purpose of achieving and maintaining a high communal rhythmic frequency. As experts in elevating and transforming energetic frequencies, they were able to energetically heal one another through the use of water. When performing such healing rituals, the Minoans would pour water into basins made out of a highly valued type of stone. They used the power of creation within their energetic minds to create energetic forces that they were able to sense and feel, and, through their unique movements, were able to successfully interact with and commandeer these energies.

Imagine a prism that is used to break light into its various color components. In the case of energetic healing, water functions in a similar way as a prism, only instead of refracting light it enhances and then re-disperses the energetic frequencies. The Minoans were able to transform and reharmonize the energetic rhythm naturally transmitted through water, thereby elevating and multiplying it into a higher energetic element that they then used for the purpose of healing. This transformed energetic element has not yet been "discovered" by modern people. The Minoans' success in performing these energetic rituals depended heavily on the harmonized bonds that existed between their minds and hearts, proving that they were true masters at maintaining thoughts of positivity within their minds and love within their hearts.

Unlike modern civilizations that have focused on the superficial aspects of water and used it only for ornamentation or recreation, the highly evolved and spiritual civilizations of the past built their sites in ways that fostered and preserved the natural movement of water. The Minoans in particular engineered various systems of transport, which included extensive stone canals and sewer pipes made of clay, to assist in maintaining the natural migration of water, permitting it to flow continuously as one.

Waterworld

As a child, I adored the rain. Almost nothing else could make me feel as happy and at peace in the universe as standing outside in a full-blown rainstorm or hearing the melodic percussion of rainfall against my window. Everything about the rain intrigued me—its perfect water particles, the refreshing breeze that often accompanied it, even the musky scent of the damp asphalt that rose from the ground afterward. At the first sound or sight of it, I would dart toward a window, press my forehead against its fogged surface, and close my eyes, as a feeling of warmth spread within my heart. I could never figure out what it was about the rain made me feel this way. What was its secret, I often wondered—the story it was keeping from me as it fell in loneliness onto the earth? It was not until the summer following my junior year of college that the rain finally decided to open up to me with its story.

After hanging out with my friend Mary one rainy evening, I was finally home and getting ready to sleep. As I lay in bed listening to the dancing rain, I directed my thoughts toward traveling to a planet I had briefly visited once before. Journeying through space at remarkable speed, I kept my eyes shut and let my energetic body direct itself. (When you finally take your own journeys through space and time, you will be blown away by the speed at which you travel!)

"Welcome. Your visit has long been anticipated," I heard a child's soft voice echo telepathically through my mind. I opened my eyes to a small, porcelain-skinned, pre-teen girl standing cheerfully in front of me. Her wide brown eyes twinkled, revealing her excitement, as light crept in from the cracks in the greenish walls surrounding us. Sporting a head of cropped brown hair and wearing a teal-colored, medieval-style cape made of a thin fabric, the girl grinned widely at me. I was about to introduce myself before she interrupted me in her gentle voice, "There's no need. I probably know more about you than you know of yourself. Come follow me!" She reached over, took my left hand in hers, and led me out of the compact, dimly lit bamboo cabin we had apparently been standing in.

Once outside, I was astonished by what I saw. We seemed to be standing on a bottomless pier made out of shiny wooden planks, and surrounding us, in every direction, was the glimmering water. All around us, I saw geometric bamboo houses built over the piers that extended out into the ocean. The vibrant foliage covering the houses glowed as the light hit it. I looked up at the sky; it was certainly light out and in fact appeared to be daytime, but oddly I could not find the sun anywhere above us. The young girl continued to lead the way across the seemingly never-ending pier that disappeared into the horizon, and as we walked, I heard the ocean lap the sides of the pier in peaceful, swooping motions.

I soon noticed a man sitting in a small wooden boat that had brown pillars or pylons holding the boat in place near the edge of the pier. Dressed in all white, his lean legs dangled over the edge and his dark, shoulder-length hair swayed in time with the movement of the ocean, covering then revealing his turquoise eyes. My curiosity growing, I walked toward him and squinted down at him in his boat. As his gaze shifted to the water, mine followed. There I could see that he was surrounded by schools of huge, exotic fish. They were all about 5 1/2 yards long, swimmingly beautifully around him in unison. All of them were mostly white, with occasional patches of bright orange splashed across their undulating bodies. They reminded me of koi fish, except that they were so huge and were shaped like eels. I could see them moving briskly through the near-transparent water in concentric swirls, and it was as though I were watching a beautiful light show.

Deeply intrigued by the fish, I decided to strike up a conversation with the man in the boat: "I just realized, I have yet to ask: What planet and dimension are we in right now?" I inquired.

The man looked up at me, widened his bright eyes, and placed the palm of his right hand lightly across his heart. "I'm sorry. I cannot tell you, as we are a peaceful planet," he replied in a tender, apologetic tone.

It wasn't long until I felt the young girl's familiar presence near me again. She gently tugged on my hand as though insisting that I leave the man alone, and then introduced herself: "My name is Eratorn." Upon hearing her words, I was overcome with a strong sense of intuition. As she stood peering up at me with her brown hair moving in the breeze, I suddenly felt that I had known her all along and that it was my incarnation into physical earth that had made me forget this.

"Eratorn, please just tell me anything you know about me and my relationship to this planet, or even the name of this planet and the dimension it's in," I pleaded with her as she held my hand.

She quickly looked down and replied just as the man had, explaining apologetically in a quiet voice, "I'm sorry. We are a peaceful planet. We cannot reveal the information."

Disappointed, I asked longingly, "Well, can't you tell me anything in relation to my very own being?"

She paused briefly before answering reluctantly, "The name we know you as is Souvwen Def."

"That's all?"

"You, along with many others, were involved in reviving the water on our planet. It was in a stage of decline, but you helped revive it."

Befuddled, I asked for clarification, "You mean you went through a famine?"

She took my hand again and we sat down together at the edge of the pier, our legs dangling freely over the water. She explained, "See, Khartika.

You cannot have thoughts of water in the physical sense. If you do, you will never come to understand its true nature. Water that exists in the energetic is very different from water in the physical. We are in the energetic presently; thus, when we speak of reviving water, we do not mean doing so in terms of how much there is. Instead, we mean its energetic functionality. Water assists in reviving the overall vibrational frequency of the planet, along with the vibrational frequency of all its inhabitants."

After pondering over her explanation for some time, I finally said, "I understand. I find I am happiest when it rains and I tend to constantly seek water when I need to meditate. I even often meditate in the bathtub!"

She nodded, moving closer to the edge of the pier so she could dip her toes into the water. "Water is one of the natural forms of true love and light. In the higher dimensions, water is part of Togetherness. It can be separated through the creation of borders and other barriers, but when it is allowed to constantly flow together, it remains united and exists as a whole, just as you see here. We do not create borders to separate or alienate it from itself; instead, we live over it in unity." I told her that a part of me had always sensed that water had the ability to multiply higher energetic frequencies, and I asked her inquisitively if this was correct. She nodded proudly, revealing her wide smile again, and replied while her feat danced away in the water below, "This is correct, and it is due to the higher energetic frequency that contributes to the formation, as well as the flow and the rhythm of water. So you do retain some information within you!"

Thinking about the different elements that existed on planet earth, I asked her, "Eratorn, what about fire and air? Do they have a purpose, too? I also know that the etheric, or the energetic, is one element that mankind has yet to discover. But it certainly is an element, isn't it?"

Upon hearing my question, she shut her eyes tightly, her face crinkling in sorrow, before answering in a sad tone, "Fire and wind, in particular, are not pertinent to the higher dimensions. However, the energies are."

I had a strong sense of what she was hinting at. "So fire only exists in the lower dimensions? Is it a creation of darkness?"

"This is very much true; however, it is not specifically a creation of darkness, but instead a creation of the natural existences on planet earth. Khartika, I'd like to show you something," she said suddenly. "Come follow me!"

She jumped up quickly to her feet and, grabbing my hand, pulled me with her across the wooden pier that extended over the expanse of the ocean. We were soon running and I felt completely weightless. Halting suddenly, she turned to me with a look of determination and exclaimed joyfully, "Now, feel it! Feel yourself as one with everything else, especially the water. But in order to do so, you must feel the love within you first. Close your eyes." I listened carefully to her instructions and, after concentrating my thoughts, I felt as though I were one with everything there. It was as though I were energetically gliding through space, and my heart echoed immense feelings of love and light.

"That is the true nature of water. It is the love that connects us all to it," she said. "I made you run because when you run, you focus only on that. Therefore, you don't have any other thoughts within your mind that could distract you and prevent you from feeling the love within yourself."

She paused, and I could tell through the intent look on her face eyes that she was telling me it was time for me to go. As usual, I was overwhelmed with many more questions that had to be answered, so I pleaded with her, holding her hand, "Wait! Wait! Please let me walk a bit more with you. Don't send me back please." She only smiled her usual beaming smile, and as we approached a group of people standing and watching the calm ocean from the pier, I could not refrain any longer. I yelled out like a wild animal, "Can anyone please tell me of my past life here? I was Souvwen Def! Please!!" No one answered; instead, they stared at me in shock. To my great embarrassment, I was sent directly back to my physical body.

Energetic Unity

In order to achieve what is often referred to as *energetic unity*, which is a connection to the higher energetic frequencies, the power of the mind and the pure nature of the heart must be in complete resonance and accord with one another. To reach this, an individual must maintain elevated thoughts of pure love and light supported by feelings of love and light. The key to doing this lies in one's ability to uplift one's mind and heart.

First, you must be able to "govern" your energetic body and remain conscious during the unconscious aspects of dreams—what psychologists and some New Agers call *lucid dreaming*. Second, you must learn how to elevate your energetic body to a very high vibrational frequency so that it can resonate in unity with the pure energetic forces of the higher-dimensional existences and their occupants. Neither of these steps can be achieved unless you have attained energetic unity. It is only after this unity has successfully been established, that you will gain the ability to roam at will through the many universes and higher-dimensional realms of endless possibilities.

It is easy and natural to travel to the lower-dimensional realms and to remain conscious in the fourth-dimensional realm, otherwise known as the human energetic realm; however, traveling to the higher realms of limitless possibilities presents a very different challenge. It is absolutely pertinent that you establish this aforementioned energetic unity, as we all exist in a universe of pure love and light. This universe does not allow people who lack feelings of love to roam aimlessly around the higher realms. As you can imagine, someone like this would be utterly detrimental to the peace and unity of the higher dimensions.

Helpful Tips

A word on thought-forms

A thought-form is an energetic creation that vibrates at the same frequency and in the same energetic rhythm as the person who created and released it. For example, if someone who is vibrating at a particularly low frequency releases a negative thought-form, that thought-form will wander around the energetic world until it finds someone else to latch onto. This person will then likely release a similar negative thought-form of his or her own, along with the thought-form he or she picked up from the original owner. What results is a domino effect, with thought-form engendering thought-form, potentially *ad infinitum*.

It's worth noting that a thought-form does not have to be negative in order for it to lower one's overall vibrational frequency. Even a more benign thought-form, such as one born out of a materialistic or superficial mind-set, can have negative repercussions. For example, if you think of yourself as a person of love and light who does not partake in the typical rat race of today's materialistic world, and instead live your life humbly and focus on your own spiritual evolution, your overall energetic frequency can still be lowered by one of these thought-forms if you're not careful. By maintaining positive thoughts, remaining alert and aware of any thought-forms attempting to latch onto you, and knowing how to shield yourself from them, you will be able to successfully avoid the consequences of allowing one of these parasites to latch onto you.

Thought-forms function like viruses. They can multiply ruthlessly and become even stronger as time goes on. If you allow a thought-form to latch onto you, and you re-release it along with your own copy of the thought-form, you may end up lowering your frequency so drastically that you eventually find yourself vibrating at the same low frequency as the majority of people living on this planet. Therefore, you must make it a priority to learn how to re-adjust your thoughts whenever a thought-form tries to gain entry into your mind. You can achieve this goal through your newfound knowledge of energetic frequencies.

For example, the thought *I am very depressed, I don't know why I feel this way* can be effectively recast as *I need to maintain a high overall vibrational frequency. I must feel this way because many people in that room were vibrating in a lower vibrational frequency.* Or similarly, re-adjust this thought, *My father is crazy; he hates me!* to something more like this: *This must be due to the different overall vibrational frequencies that we vibrate in. Therefore, I must not get so frustrated.*

The rain

In the higher-dimensional realms, rain is a natural rhythmic force that assists in the maintenance of the universal cycle of higher vibrational frequencies. Rain plays a slightly different role in the lower dimensions, including right here on planet earth, where it contributes to the elevation of the energetic rhythmic frequencies within individual people.

A lot of people say that they sleep better when it rains and often find themselves dozing off when it begins to rain. Although this may sound odd, such a reaction to rain is only natural. When we sleep while it is raining, our energetic body vibrates in a frequency that is significantly higher than normal. Our energetic body is automatically, although only temporarily, pulled to a much higher-dimensional existence than it would normally occupy, the fourth-dimensional realm of the dream dimension. Because these higher dimensions vibrate at an elevated frequency, our energetic bodies are consequently reenergized and recharged to the purest forms of the natural forces of love and light.

When we wake up, our energetic body continues to temporarily resonate at the same frequency it does in the higher-dimensional realms; this is precisely why many people report feeling especially great when in rains. As well, rain assists in blocking unnecessary and energetically harmful electromagnetic forces.

Lightning

I am only allowed to touch briefly on the concept of lightning in this first book. Lightning is essentially a collection or amalgamation of the concentrated focal components of the force of rain that we see in the physical world. Lightning occurs at virtually the same time in the energetic world as it does in the physical world, energetically manifesting itself into the physical in less than a split second. In the physical world, the frightening sound of thunder tends to accompany rain; however, thunder is merely a higher frequency or resonance that has been forced to manifest itself in a lower-dimensional nature.

On various occasions, people have reported gaining special or paranormal abilities after being struck by lightning. The explanation behind this enigma lies in lightning's functionality as a concentrated, focal energetic force of rain. Lightning carries with it resonating frequencies that typically exist solely in the energetic sphere of the universe. Therefore, when someone is struck by lightning, he or she is actually being "struck" by a vibrational frequency that is higher than that of the human energetic body. Due to this temporary increase in vibrational frequency, people often become more sensitive and attuned to the energetic world. There is still much research to be done on the nature of lightning, but one thing remains clear: If we are striving to truly understand how changes to the energetic body can cause alterations in the physical, we should remain fervently focused on studying rain and its relationship to lightning.

The Water Exercise

If there is an element that voices its rhythm in a concise and clear energetic manner, that element is water. This exercise will enable you to align yourself with the high, elevated rhythm of water and further enhance and maintain your overall vibrational frequency. You will need to be in close proximity to water, whether it's water in its natural environment, such as a lake or stream, or the water that you've just used to fill up your kitchen sink. Success in this exercise does not rest on being immersed in water, but rather on your ability to dedicate your thoughts to the rhythmic capabilities of water. You must not depend merely on the water itself for achieving the aim of this exercise, but also on the innate power that exists within your mind.

To begin, fill a basin, sand bucket, or bathtub with water. Any kind of container will do.

With your eyes closed, slowly submerge the tip of each finger, one following another, into the water.

As you put your fingers in the water, envision a luminous, blue light moving from the tip of each finger, through your hand, through your arm, and then dispersing itself throughout your entire being and forming an interconnectedness between the water and your energetic self. (In the energetic world, the color blue is associated with the energetic essence of water itself, and hence has the ability to both heighten and magnify vibrational frequencies.) You must feel the cooling sensation of this energetic, blue light in order to maximize the effect of this visualization.

Follow with the other hand while keeping your fingers in the water and dedicating your thoughts to water's ability to heighten your energetic frequency.

5

Deception

The crisp afternoon air collided with the bird's jagged ebony wings as he soared aimlessly through the sky. He had strayed from his usual path today; perhaps he had wanted to venture into the lands of sunlight and lush trees, just for a change, far from the dark and grim province he had always dwelled in. As he glided haphazardly through this very different world, he caught beneath him the sight of a society of birds that was entirely unfamiliar to him.

Upon his descent, they welcomed him with their usual hospitality and immediately took him in as one of their own. Their society boasted an array of different species of birds, yet he was unlike any other they had seen before. They all possessed wings, just like the new visitor, but, ironically, had yet to discover how to use them. Their wings had another purpose, though: They used them for collecting endless assortments of flowers, colored pebbles, and rocks from their vibrant forest to serve as ornaments. They took great dedication and care in embellishing themselves with whatever they gathered, and used such decorations to flaunt their wealth and stand out from all the other bird societies. It was their way to be noticed, but it came with a heavy burden. They either didn't know or possibly didn't care that placing such weighty ornaments on their wings hindered them from realizing their ability to fly.

After spending a day with his new acquaintances, the visiting bird had won over more than just their hearts; he had become the victor of their minds. With his beady black eyes, he glared at the society's members with selfish desire, as he imagined how easy and empowering it would be to rule over such a gullible society. Dazzled by his presence and charm, the central power of the society bestowed their trust, appreciation, and respect on the visitor, and the very next sunrise admitted him into the community. That same day, greedily celebrating his success, the dark bird set out to create his agenda. He quickly came to the conclusion that the best way for him to rule and gain unsurpassable control over this society would be by continuing to keep its citizens blind to their true abilities. He quickly created a manifesto of false information and sought a candidate to become his puppet ruler. He recognized immediately that the best candidate for the job would be the bird with the weakest mind yet also one that secretly possessed an intense desire to rule over others. He wanted to select the bird that saw eye-to-eye with him.

It was not long until the dark bird found the ideal candidate for the position, and deep within the dark underground caves below the vibrant forest, he informed this bird of his own ability to fly. He revealed the least amount of information possible, for he knew better than to teach the bird of the true knowledge that would enable him to fly. Instead, he took great care to infiltrate the chosen bird's mind with lies and misinformation, recognizing that, when given the opportunity to rule over others, the bird would efficiently disseminate this information in a way that would benefit him.

Many sunsets passed, and the once-lively society of embellished birds and lush greenery was replaced by rotting trees and infiltrated minds. The talk of the forest was now of the new ruler's ability to fly, an ability that did not apply to the rest of them, for according to the ruler himself, he was the only "special" one.

Every person's innate ability to travel multidimensionally comes with the great responsibility of learning to discern between the knowledge he or she receives from various sources. Our universe is one of limitless possibilities, a world composed of endless puzzle pieces, many of which have yet to be found. Anyone who claims to be special or in possession of a unique ability that you cannot achieve is, without a doubt, lying. There is absolutely nothing in this universe that someone else has done that you cannot also achieve yourself, for we are *all* beings of limitless abilities. Nevertheless, it is crucial that we do not simply accept those pieces of the puzzle that have to be forced into the empty spaces just to get them to fit. We all aspire to catch any glimpse of the big picture—that is, truth—that we can get, but the beauty of understanding this mysterious universe comes through solving it ourselves. It may be an arduous puzzle to solve, but it is unquestionably the most spiritually rewarding one.

Deception

Deception can occur in a variety of different forms in the energetic world. Even in the physical world, it can throw you off your destined path. On earth itself, there exist numerous people who blindly receive information from sources they remain oblivious to and ignorant of; yet without even thinking twice, they then spread this information to those who actively question the multidimensional existences. Therefore it is absolutely crucial that you learn how to use discernment with the information you receive during your multidimensional experiences. Despite the best of intentions and wanting only to assist your fellow human beings, if you do not carefully discern among all the information that you are given before sharing it with others, you risk creating more obstacles for the spiritual evolution of humanity and forming clouds that further hinder people from seeing the clear picture of reality and truth.

The first form of deception

The first form of deception includes lies that are disseminated by beings of the lower-dimensional energetic realms. Remember that when you embark on a journey in a negative mindset and with negative emotions, it is almost guaranteed that you will end up traveling to the lower-dimensional worlds. Unfortunately, I was not always aware of this. In my earlier travels, I was often overcome with great fear when I would find myself in terrifying situations. Sometimes lower-dimensional beings transferred their electromagnetic resonances to my tech gadgets and appliances. At night, I would receive numerous phone calls from family members and friends in my contact list, who would later claim that they had never contacted me. I became so overwhelmed with terror that I soon began to sleep with all the lights on in my room.

I also had yet to come up with the appropriate protocols for traveling pleasantly and safely to other dimensions. As a consequence, all of my early journeys were accidental and generally "automated" in nature. Often I would find myself in complete darkness within the confines of my own room, entirely unable to open my eyes. My body would be overcome by a heaviness of such intensity that I was unable to move, as though I were paralyzed, and I would hear beings approach me and telepathically communicate with me in distinct voices that I soon became all too familiar with.

They would introduce themselves to me as spirit guides; some even claimed to be angels. One particular entity that I regularly communicated with in my earlier travels was able to deceive me by claiming that she had previously lived as a human on earth before dying and becoming an angel. She encouraged me to focus my meditations on her, instead of on the force of universal love and light, as I had been, and in return promised that she would always be there for me in times of need, unlike the light beings who had visited me once and never returned.

At the time, I was a young, naïve, and vulnerable girl, and I blindly believed every word this being spoke. I would soon find myself terribly misled by her. With her beguiling encouragement, I made the choice of focusing my meditation on her instead of the goodness in the universe. I continued conversing with her in the same paralyzed state every night for the next three weeks, and it was only at the end of those three weeks, empowered by feelings of love and light, that I finally was able to see her true appearance. That particular night, I concentrated my thoughts on visiting her, and although I was vibrating in a much higher frequency than before, I was strangely able to reach her lower-dimensional level of existence instead of the higher-dimensional existences I normally would have been drawn to. It was then that I truly saw her for the first time. She was a gruesome being of absolute ugliness and evil, closely resembling the evil spirits typically depicted in horror movies. As she appeared in front of me, I received the firm intuition that she had indeed previously been a human being who had later become a lower-dimensional being as a result of the hatred and negative emotions she possessed.

With all the cards finally laid out on the table, the extent of her deception was revealed to me that night; but because I was vibrating at a high frequency and my thoughts and emotions were focused solely on love and light, I simply said to her, "I pity you. I send you my love and light." As soon as the words left my mouth and she recognized the love within me, she began to run amuck and scream in great agony, as though my words were torturing her, until eventually she—it?—dissipated into absolute nothingness.

This particular experience led me to my first discovery of how the positive forces of love and light can be used for two purposes: one, for shunning negative beings; and two, for preventing yourself from slipping into the lower-dimensional existences when you are sleeping. I learned that holding onto even the slightest negative emotion upon falling asleep would virtually guarantee a trip to the lower-dimensional existences. No matter how unpleasant it may be to accept reality, the truth is that there exist many dark beings within the energetic realms that will attempt to either inflict fear upon you or throw you off your path to discovering the truth about the universe and your boundless potential. Wherever you may find yourself in this universe, you must never succumb to forces of such darkness and negativity, for you are a being of love and light, and it is only through uncovering the truth that you will be able to sharpen your limitless abilities.

The second form of deception

The second form of deception usually occurs when you are exhausted or half asleep. After opening yourself to the world of multidimensional realities, you will surely come across numerous enigmas and confusing yet enticing situations that you will want to explore for the sole purpose of satisfying your curiosity. Even if you are in a positive state of mind, physical exhaustion can automatically cause you to vibrate at a lower frequency. This occurs because we are programmed that way: As long we reside in our physical bodies, the connection between the energetic and physical cannot be severed. The two will remain interconnected with one another until physical death.

My first experience with this particular type of deception in the energetic world occurred while I was on a trip to India. For reasons unknown, there exist numerous energetic portals throughout this country. On my second night there, utterly exhausted due to the unavoidable jetlag, I was falling asleep and slowly detaching from my physical body when I noticed a small red door with a peculiar doorknob made of jet crystal manifest on the ceiling above me. It slowly opened to reveal one of my good friends, who was supposed to be in London at the time. She smiled cheerfully at me, her white teeth glimmering, as she waved at me to come through the door with her. At first I was reluctant to do so, but that thought disappeared as soon as I heard her recognizable voice resonate through my mind, "Come! Isn't it exciting? Don't you want to go on a spiritual adventure together?" The familiar voice immediately put me at ease, and I let my guard down. *She's one of my greatest friends, after all,* I thought to myself. I lifted myself using the force of sheer willpower and slipped through the door alongside her, only to find myself in a ruby-red, round tunnel encompassed in what looked like intertwined weeds made of metal. I instantly felt that something was off, for even in this early stage of my multidimensional experiences, I was well aware that different colors depicted different spiritual vibrational frequencies in the energetic realms. I had learned that red, in particular, was considered to be a negative color in the energetic sense. I stopped following my friend as she continued to move through the dimly lit tunnel, and quickly turned back, nervously searching for the red door we had entered from. When, with relief, I spotted it, I felt an excruciating magnetic force pulling me from the back, and I knew instantaneously that it was the being that had appeared to me as my friend. The electromagnetic force shot me out toward the top of the tunnel, and I landed on the ground with an excruciating thump. I immediately began ordering my mind to wake myself up, as I knew that this would be the fastest way to escape the attack I was experiencing. Fortunately, within a few seconds (which had then felt like hours) I found myself back in my physical body.

When traveling energetically, you are almost always guaranteed a safe return as long as your physical brain remains continuously replenished with oxygen—in other words, as long as you remain alive in the physical sense. In light of this, whenever you find yourself in danger while traveling, the fastest way to return to your body is by alerting your brain. Each and every one of us has this gift that will bring us back swiftly into the physical whenever fear or danger are present. This gift functions like an invisible safety net, and although it is good to rely on it and trust it, in order to gain full access to the energetic realms without any disturbances, you must try to put the emotion of fear aside.

Beings of the lower-dimensional realms can alter their energetic appearance for the purpose of deceiving you. They are also capable of strategically directing frequencies to your brain, enabling them to gain access to familiar voices stored within your long-term memory, which they can then use to deceive you even further. So, always trust in your senses, and do not blindly follow every being you encounter in your multidimensional travels.

The third form of deception: Jatakawarta

On a warm, humid August night, I went on a particularly memorable journey. It marked the very first time that I was able to remember the name of the planet I traveled to after returning to my physical body. From my many multidimensional adventures, I had acquired ample experience with the standard protocols of multidimensional travel (see the Appendix for a short list of these protocols), but only recently had I learned of its fundamental aspect, particularly the ability to navigate via the governance of the mind. When you're traveling multidimensionally, it's not unusual for your subconscious thoughts to suppress your memories of these travels. Most of the time, rather than being able to clearly recollect everything that happened, you'll have collective memories in the form of symbols, which will then manifest as dreams within your subconscious mind. Thus, maintaining complete control of your subconscious mind becomes absolutely critical in being able to travel freely by will and having full recall of the events.

By this point in my life, I had already learned and understood that the power of the heart was just as important as the power of the mind. I traveled far above and beyond earth until it became merely a faint recollection, a tiny speck in the universe. With no particular destination in mind, I simply allowed my intuition to lead me, concentrating on the feeling of love as I traversed through the emptiness of space. Every now and then, I came across an energetic planet I had never seen before, each one entirely unique. There were thousands of them, and as I peered at them from a distance, I shook my head at the thought of the paltry nine that had been discovered by the scientists of our planet. To this day, I can vividly recall the feeling of pure excitement that overwhelmed me at that very moment. Never before or since have I felt as much freedom as I did on that day!

As I continued traveling freely and aimlessly through the universe, I finally came across a planet that strongly reminded me of planet earth. Its resemblance to earth was striking; the only difference was that this particular planet was entirely composed of luscious greenery on one side and clusters of breathtaking oceans on the other. Having always felt a deep attraction to the ocean, I was automatically drawn toward the oceans and found myself flying to them at great speed. I took a deep breath of the exhilarating, fresh air and stared longingly into the expanse of the shimmering blue water. My gaze soon came across a curious yet beautiful island covered entirely in iridescent white sand. On it were five majestic pyramids, blinding gold in color. I marveled at the precise placement of these pyramids, four skillfully positioned at the corners of a perfect square, with the largest pyramid at the center.

There was another, similar island, also entirely covered in the same twinkling white sands, but this one was oval shaped. On this island, I could see a golden pyramid with two majestic pillars that met at its highest point. Utterly awestruck by this wonder, my concentration quickly shifted when I caught sight of a man in the distance. With a shining golden staff in his left hand and a furious frown on his face, he galloped toward me on a white stallion. As he drew closer, the glare of his chunky golden armor forced me to blink a few times. He reminded me of a warrior during medieval times, the kind that would have appeared in battle alongside King Arthur. I could tell he was coming fast, the white sand shooting out into the air behind his horse as it careened across the island dunes. It instantly became clear to me that my presence had been misinterpreted as a threat, and so I made my way to the other side of the planet.

I was initially astounded by the green side of the planet, particularly by its marked resemblance to certain impoverished regions of India. I found myself in what seemed to be a deserted village where muddy houses—or, more accurately, unsteady shacks—were arranged in a single crescent that was entirely surrounded by wide, luscious forests. Each shack was composed of two levels, with a makeshift balcony protruding from the higher level. As I made my way through this curious town, I noticed a woman dressed in a mustard-yellow Indian sari sitting cross-legged on the balcony of one of these houses. There was an enormous, moon-shaped stone bowl in her lap, and as she muttered various mantras underneath her breath and placed her hands over the bowl, a fire ignited inside of it. Although she was seated, I could tell that she was a few inches shorter than the average human, and I wondered at her disproportionally large head.

I continued walking, my gaze bouncing from one dirty house to another, until I reached the last house. With its doors wide open, it seemed to be welcoming me inside. Led by my intense inquisitiveness, I entered the house and immediately stumbled upon a diminutive man. He had deeply tanned skin and short brown hair, and looked as though he was of Indian or Pakistani descent. With a bewildered look on his face, he stared at me in silence for a while, before asking telepathically whether I needed assistance. I asked him whether this planet was earth. "No! This is certainly not earth! Most certainly not!" he responded, with an air of arrogance, implying by his tone that earth was a place for backward or less-evolved beings. When I asked him where I was, he replied, "This is *Jatakawarta!* Though we exist in our physical, too, we are *far* more advanced than residents of earth! How else can we communicate and see you even when we are in our physical bodies? Earth! How could you think of this as earth?"

I repeated the word *Jatakawarta* a few times in my mind, ingraining it into my memory so that I would remember it upon returning to my physical body. I asked him if he would kindly tell me more about *Jatakawarta.*

"This is a place during Ashoka time. Ashoka Period, we call it. You must know Ashoka," he stated, with a demanding look of certainty on his face. He then continued, squinting his eyes at me, "Earth...do you know that we can come and go to earth at any time? We came through the caves of the Himalayas. We also often use the Nepali border region. No one notices us when we walk amongst the humans. Humans—they fear the unknown; they fear ones who are different. So they choose not to notice differences when the difference is small." I telepathically agreed with him and nodded my head, thinking to myself how unusually large his head was. He quickly caught on to my thoughts and proceeded to laugh, before asking me with a puzzled look whether I was from earth, for he could not comprehend how I was able to visit his planet if this was the case. According to him, energetic travelers no longer come from earth due to the various restrictions and barriers that have been put in place to prevent them from traveling interdimensionally.

I glanced at him with a frown after hearing his explanation and asked, "What is the current year, if you may tell me?"

His face was suddenly overcome with an injured expression, as though the concept of time itself was somehow an impertinence or personal insult. "I've told you: This is Ashoka time! You said you've understood! I would've thought of you as more advanced than those backward humans. It is Ashoka time!"

Hoping to not upset the man with my presence or questions any longer, I asked him to tell me of any other nearby planets or continents with the same frequency as *Jatakawarta* that I could travel to. He mentioned a nearby planet whose "rotational frequency" (in his words) greatly differed from that of *Jatakawarta*. After thanking him for his hospitality, I was soon on my way.

I crouched down to exit through the small back door of the house and began to focus on the usual protocols for directing myself to my next destination. I was soon distracted when I heard a cacophony of screaming voices growing louder and louder: "Catch her! Catch her! Catch her!" The crazed words hit me like waves of a violent storm, and primal fear instantly kicked in. I began to lift myself upward using the power of my mind, but as I ascended, I felt as though I were being suffocated. It was not until I glanced down in frantic confusion that I noticed someone's arm wrapped firmly around my rib cage. I felt the arm's grasp gradually tighten, and soon a scorching sensation of pain shot through my body; it was as though the life-force were being pulled out of me. Twitching from the agonizing sensation, I quickly looked over my shoulder in search of the source of the pain, only to find the same woman I had seen earlier on the balcony of the house, staring sickeningly at me. I telepathically ordered her to let me go, to which she viciously declared that I was hers to keep.

As I squirmed violently to free myself from her tenacious grasp, I received a message through an accidental telepathic linkage between our minds, informing me that she wanted to keep me caged like a wild, exotic pet, as she believed that there were many benefits to owning an interdimensional traveler. Because different frequencies hold different means of knowledge, achieving complete control over an interdimensional traveler such as me would enable her to gain knowledge and power for her own selfish purposes.

As I considered this horrifying thought, I felt the life-force being constricted out of me. Although I was in my energetic form, I had thankfully managed to maintain a link with my physical brain during my journey. Helpless under the woman's immensely powerful grasp, I began to directly transmit energetic thought waves from my energetic body to my physical mind. These thought waves carried the message that I was in danger and begged my brain to physically wake me up so that my energetic body could be pulled back into my physical body. I abruptly (and with great relief) woke up in my room feeling extremely exhausted and drained. Though I was back in my physical body, I continued to feel the same painful sensation for the next 15 minutes. This was due to the last vestiges of the link between my energetic body and my physical mind.

Following this incident, I came across various works of Sanskrit literature, particularly the great epic stories that are collectively known as the *Legends of Ashoka*. I discovered that Ashoka was a mythical king who had apparently enforced Buddhism as a religion. A number of these Sanskrit illustrations as well as writings were collectively known as the *Jataka Tales*, referring to Buddhist tales that largely spoke of the various *bodhisattvas*, or mythical enlightened beings, as portrayed by the

Buddhists. After studying the ancient writings in detail, I further uncovered that *warta* signified "city." Thus, my travel had in fact led me to Jataka-city, or the "story-city." It became clear to me that the beings of *Jatakawarta* had once possessed strong energetic connections with earth's past civilizations, and it is highly probable that they openly visited and communicated with these previous civilizations. Considering the current problems in our universe, we will unfortunately never know whether or not their realm was overcome with darkness and as a result coerced into treating visitors in such negative ways.

From my experience in *Jatakawarta*, I discovered the truth to the enigma of time as a function of space and distance. The human concept of time does nothing more than confine the human mind. If people were instead taught to understand time in the context of distance, instead of continuing to blindly accept the idea of time as a constraint, limitless journeys to various realms would be possible for everyone. But perhaps the most valuable lesson the universe taught me from this particular experience is the importance of knowing how to differentiate between positive and negative beings. Even in our current, "advanced" society, it is entirely possible to encounter negative beings due to the darkness that has ruthlessly spread through the various realms. It is best to learn how to recognize these beings, rather than to risk placing yourself in dangerous situations.

The fourth form of deception

The fourth type of deception occurs due to our inability to discern what is essentially a theft of energy. Many people visit religious sites and sacred locales in an attempt to pay homage to the deity or deities of their respective religion. It is with absolute blindness to the energetic world that such people continue to follow their religious rituals without using their own ability to sense and discern whether such actions are energetically beneficial or not. By way of an example of this, through one particular multidimensional experience, I came to discover the true meaning behind certain religious statues found in temples, sacred sites, and the like.

When my two friends and I were in Kyoto, Japan, we decided we wanted to visit a temple. Following my friends as they strolled in front of me, the stiff wooden floors of the temple creaked under each step. Earlier that day, we had been told by the courteous hotel staff that this was one of Japan's most prominent temples, with hundreds of impressive statues of guardian deities carefully placed throughout its grounds. There was nobody else around, aside from a hunched-over janitor who was meticulously mopping the wooden floors.

As I stood at the end of a long, dimly lit hall, I could smell the distinct scent of sandalwood, and it began to constrict my breathing. (I should have known by then that my asthma and visiting temples was not the best combination, but I was always up for a good adventure.) My friends rushed toward the wooden donation box in the middle of the hall, eager for a chance to pray to each guardian deity statue so that they could proudly check off another box on their cultural bucket list for Japan. They each dropped a few coins into the small slot on the top of the box. Clasping their hands together, they recited a short mental prayer, and then continued on to the next statue as they moved toward the exit. I was running out of money, as I had purchased pricy Buddhist praying beads for my father earlier that day. Placing my hand over my right pocket, I could feel at least 100 yen through the thick denim of my jeans—just enough to pray to one deity. With my friends well ahead of me, I quietly tip-toed to the center of the hallway, dropped my 100 yen in the wooden donation box, and grabbed an incense stick.

As I concentrated on lighting my incense, through the corner of my eye I suddenly noticed a swift movement near the grouped statues. I could have sworn I had seen something move and quickly looked over my left shoulder to see if my two friends had noticed it, too, but they were both focused on their praying. I curiously gazed at the deity they stood in front of and concentrated on its structure, until I felt a sickening pang of negativity resonate through me. The temple suddenly became a blur, and everything around me began to spin. I immediately turned back and rushed out of the temple, hurrying away from it until I collapsed on a cold marble bench and inhaled a deep breathe of Kyoto air.

I soon saw my friends sprinting toward me from the exit of the grand temple, their expressions stricken with concern. Placing their hands on my forehead to feel for a fever, they asked me worriedly what had gotten into me, but I couldn't adequately explain what had happened, other than a feeling of exhaustion and constriction had come over me. I asked them if we could go back to the hotel, and they nodded in agreement, handing me a can of club soda. I have no recollection of what happened once we got back to the room, but my friends would later fill me in.

I immediately felt myself being automatically detached from my physical body by a magnetizing force, and I was instantly surrounded by a collective noise of sinister laughs and frightening voices. I knew right away that I had been brought into a lower-dimensional existence. After scouting out my grim surroundings, I soon realized that I was in the energetic locale of the same temple grounds I had visited with my friends earlier that day. Everything around me was entirely dim, colorless, and lacking in life. While the temple structure itself looked the same, the statues now had grotesque beings occupying them.

I observed a group of boisterous schoolchildren enter the temple in their physical bodies. After hearing the instructions of an exasperated teacher, they finally stopped their chattering and stomping, and settled next to the box of incense sticks in front of the copper statues. I could clearly make out the prayers that they directed from their minds. One boy, with his eyes glued to a particularly fierce-looking statue, vehemently prayed for his drawing to win the school award; another, wearing thick-framed glasses and with his hands clasped in prayer, politely asked for his sister to be accepted into her university of choice. As they silently stood there, with the crisp, familiar scent of incense spreading through the temple, I noticed orbs of bright blue light of increasing brightness begin to manifest above the children's heads. Enthralled by their shimmering glow, I followed them until my gaze fell upon the statues. The orbs swiftly disappeared into each statue, while the grotesque beings perched inside greedily attracted and absorbed the blue light; in return, the beings emitted a haze of murky, brown clouds at the children as soon as they finished their prayers.

As I watched in confusion, I was suddenly empowered by a thought: These soulless, lower-dimensional beings that possess nothing but energetic bodies vibrating at incredibly low frequencies are able to appear and exist in the human energetic dimensional realm (higher than their own) by consistently stealing the vibrational frequencies of humans—in this case, children. An individual's vibrational frequency is at its highest during the act of prayer, or when his or her mind and heart are perfectly aligned. Taking complete advantage of this knowledge, these lower-dimensional beings ensconced themselves inside the metal statues of the temple and tainted them with their frequencies so that they could directly transfer those energetic frequencies to people. They were well aware that by residing within a statue that was made of metal, particularly copper, they would be able to attract and absorb the positive rhythmic vibrational frequencies transferred from the children through prayers and feelings of positivity, in order to help elevate their own. Just as metal magnifies sound, it also retains and magnifies rhythmic frequencies long after receiving energies transferred from people. This ability to "redirect" energies enables these lower-dimensional beings to remain in the fourth-dimensional human energetic realm longer than they would have otherwise. Unfortunately, a few of these vile being soon spotted me. As they eased out of the glimmering metal statues and began to approach me, I immediately concentrated my thoughts on getting back to my physical body by alerting my physical mind to wake itself up.

When I finally managed to prop my eyes open, I found myself awake in my physical body and practically glued to the hotel's signature orthopedic bed. With my thoughts (and sheets) scattered in every possible direction, I was soon brought back into the present by my friend: "You do realize that the moment we opened the door to the room, you fainted? Were you dehydrated from walking all day in the scorching sun?" she asked as she stood over me, struggling to unwrap the convenience store *onigiri* that she had instantly become enthralled with upon our arrival in Japan.

"No, I'm fine," I replied to her with a reassuring shake of my head. As I forced myself to get out of bed to brew myself a fresh cup of green tea, I heard my other friend remark matter-of-factly, with her brow arched in confusion, "It's really rather strange. I don't know why, but I also felt so tired and drained after we went to that temple. I wonder what's wrong with me. It must have been the heat or something." She shrugged it off and proceeded to get ready for the day's upcoming adventures, but I knew that I had the answer to her question embedded in the recesses of my mind. No matter how much I wanted to share it with them, I had to keep myself from blabbing it to my two logically minded, valedictorian friends who reminded me of the Vulcan race from *Star Trek*.

What they did not know, I had, in fact, discovered during my temple journey: When someone prays to metal statues at sacred sites such as the one we had visited in Japan, his or her vibrational frequency is actually lowered, not elevated. Therefore, instead of seeking sacred grounds as outlets for prayer and supplication, you should focus on reconnecting yourself with sites in nature that bear the element of water in order to elevate your own vibrational state of frequency and share your wishes with the universe. After all, the true higher beings that these statues portray exist in the universe as formless energetic beings of incredibly high vibrational frequencies. They do not actually bear wings, grow beards, or even appear with swords, as we have been conditioned to imagine them. In the case that you still find it necessary to go to such sites to pray or elevate your vibrational frequency, you must remember to use your tools of discernment—sensing and feeling—when faced with the task of choosing the correct place to do so.

The fifth form of deception

The fifth type of deception also occurs in the human dream realm, otherwise known as the fourth-dimensional existence. It is within this dream realm that humans translate their automated thoughts—that inner "chatter" or running commentary that most of us contend with on a daily basis—into what we know as subconscious dreams. Deception in this dimensional plain normally occurs either at the very beginning of a journey or starts right in the middle of it.

After acquiring the fundamental knowledge of multidimensional travel, you will automatically begin to vibrate at a significantly higher frequency, and this rhythm will be heard and identified by beings of all dimensional existences. Because the actions of negative beings of the lower-dimensional realms are motivated primarily by jealousy, they will tell lies to get you to believe that you are communicating with beings of the higher-dimensional planes, while keeping you trapped inside the dream realm.

My encounter with the three light beings that I discussed at the beginning of this book marked both the first and last time these beings visited me. I never saw them again, not even during my most frightening multidimensional experiences. Their absence perplexed me and in fact made me wonder if they wished for me to learn everything myself the hard way.

It was on the night before my first final exam of my senior year in college that I learned of the dangers that come with allowing the mind to remain inattentive in the unconscious state during sleep. I initially did not think that I would travel that night, expecting instead to spend my sleep time roaming around the human dream realm, as I was well aware that I did not have enough "fuel" to maintain a high vibrational frequency on such a stressful day.

I drifted into a deep sleep and soon began to have a fairly normal subconscious dream of sitting in an examination room and then failing my exams. With my mind in an automated state and creating situations related to my daily subconscious worries, I continued on to a vivid scene in which I witnessed myself shamefully graduating without honors. Everything that happened in this dream felt all too real, until a ray of bright light suddenly blinded me and three light beings appeared in front of me.

Image of a manifestation occurring in the human energetic realm.

"Why are you dreaming when you can instead stay alert?" one of the beings telepathically communicated to me. Surprised by their return, I quickly explained, "I am tired. It takes energy, you know." I paused, and then continued, "Why have you not visited me all this while? Do you not know of all the difficulties I've had to face?" They seemed to completely ignore my question. Then, one of the beings hastily approached me and demanded, "We do not have time; get out of this automated dream realm! We just have to inform you that one of our kind has gone astray and turned dangerous. Do not communicate, and stay safe! We will open a portal for you; go through it and you will be safe there." Upon hearing these directives, I naively figured that it would be best for me to listen, and without considering the situation any further, I agreed. I caught sight of a cloud of purple energy in front of me—the portal, no doubt—and went through it as fast as I could.

To my horror, I found myself in an entirely different existence. I appeared to be in some kind of compound positioned in the middle of immensely tall jagged rocks that encircled me. As I frantically looked around, it reminded me of Stonehenge in England, except that it had an incredibly negative, hostile energy. I moved closer to the enormous rocks that towered threateningly over and around me, but when I tried to climb around them and escape, I was violently thrown back onto the ground by a powerful, unseen electromagnetic force. Obviously there was some sort of invisible barrier. I tried countless times to escape the cruel confinement, only to be repelled back onto the hard ground with a loud and painful thump.

A girl's quiet, forlorn voice unexpectedly interrupted my efforts: "You cannot leave. Your consciousness has been trapped within this energetic compound, and your physical body will most certainly enter a coma that will lead to death after an earth time of a few months has passed. You fell into the trap. I've been here forever." At first, I could not seem to find her, but after turning around I found a frail girl sitting hunched over a few feet away from one of the sharp rocks. She looked to be in her early 20s and was dressed in a light blue, one-piece jumper. With her downcast eyes peering through her stringy blonde fringe, she let out a sigh before whispering, "It's a trap. Can't you see? Anyone who deviates from the dark force and is a threat to them is placed here. They can send stimuli to your conscious mind to alter your perception of their energetic resonance. So they have the technology for altering their rhythmic frequencies to the same level as whomever they wish to mimic. Everything they did was an effort to cause a coma within your physical mind and trap your energetic body here for as long as they can. They see you as a threat." She tilted her head slightly backward, and her grayish-blue eyes peered out from underneath her hair to look at me: "You're a human, right?" she asked inquisitively.

"Yes, I am a human," I answered.

"I had a life as a human once. It was not so long ago. In fact, it was before this current life of mine, in a place known as America. I remember it being quite fun."

Although I desperately wanted to believe her so that I could finally figure out where I was, I did not have the slightest idea of what to make of her. I had no sense or intuition regarding her nature or intentions. For all I knew, she could have very well been a projection created by the same dark beings that had tricked me into thinking they were the light beings. I thought that she could quite easily be an empty vessel completely void of life-force and with an energetic body completely overtaken by the dark forces.

After one final failed attempt to break through the invisible barrier that kept us trapped, I sank to the ground and focused my thoughts on communicating with the higher beings to ask for their help. Time seemed to fly by as I attempted to reach them through intense concentration, until I felt my own sense resonate this knowledge through my mind: "Remember: Everything in the energetic world is created through thoughts of love and positivity, and energies can once again be rearranged through thoughts of love and positivity. You are in control." Upon hearing this, I immediately closed my eyes and envisioned myself trapped inside a room by an energetic shield. I began to imagine myself as a rhythmic vibrational field of energies and centralized my thoughts purely on the forces of love and light. I felt the resonance of these thoughts begin to vibrate through my energetic body, and I was soon lifted through the invisible electromagnetic barrier that had ruthlessly kept me a prisoner, until I abruptly found myself back in my physical body.

"Oh. My. Goodness," my mother gasped in her usual dramatic manner as I unsuccessfully attempted to scurry through the living room without being noticed. It was already dark outside, and more than three hours of earth time had passed. My mother was having one of her usual dinner get-togethers with family members and friends, and she had successfully diverted just about everyone's attention to me. As I stood in the middle of the living room in my favorite pajamas, gazing blankly around me, my confusion must have been palpable. My younger sister blurted out, "You've been asleep since 2 p.m. yesterday and it's currently 10 p.m. today. Are you okay? All of us were considering waking you up, but it's good that you're finally awake now." Her eyebrows raised in expectation of a response, but I simply mumbled, "Yes, I'm all right," and made my way to the kitchen to make myself a cup of coffee.

As I stood in the brightly lit kitchen, leaning against the cool granite counter and twirling my spoon inside my cup, a smile spread over my face. The higher beings had never actually stopped communicating with me! They had given me a kind of energetic phone line through which I could safely communicate with them through my own senses and feelings in the presence of love and light. They had remained alongside me throughout my entire journey.

A Word on Psychics

Most of us have seen the flashing neon-letters reading "PSYCHIC" or "TAROT CARD READING" plastered on the windows of small, creepy-looking establishments. But what exactly is a psychic, or a medium? In reality these are just titles, used by people just like you and me who have managed to learn how to resonate at the same vibrational frequency as the energetic beings with whom they communicate. As you may have already guessed, the unfortunate truth is that the methods that these people use to become psychics and mediums only end up discrediting them in the end.

Psychics in ancient civilizations served as teachers and advisors to those who wished to pursue their studies through esoteric means. The majority of them lived their lives with a deep gratitude for the presence of love and light, and this enabled them to communicate with beings of the higher realms through their sense and intuition. Most of the so-called spiritual mediums of today are restricted to communicating solely with earthbounds or lower-dimensional beings. After attaining a trance state, these individuals remain physically awake while their minds focus on altering wavelengths so that they can communicate with nearby beings. This can be dangerous, as they are essentially opening themselves up to communicating with negative beings in the fourth-dimensional world.

The human mind operates in an incredibly similar manner to an antenna, receiving wavelengths and frequencies and translating them into sounds and images. However, no matter how long a psychic remains in his or her physical body while receiving these frequencies, his or her energetic body will continue to reside within the human energetic realm. The laws of the universe make it impossible to remain conscious in the physical body while traveling energetically to the higher realms. This is because the human energetic body automatically functions in the fourth-dimensional realm when it is awake. Similarly, beings that occupy the higher-dimensional existence cannot lower their frequencies to a much denser state simply for the purposes of helping you figure out whether Billy is indeed your soul mate.

When these spiritual mediums succeed in their ventures and are able to provide you with accurate information, this means that they have achieved successful communication with your energetic body in the energetic dimensional realm. Even if a psychic has managed to provide you with the information you asked for, it is still a bit silly to consult with one. If you've not yet sharpened your ability to stay in synchrony with your energetic body, it may make sense for you to seek out a medium during desperate times. Instead, why not brush up on your own abilities by speaking to your energetic body, which bears an endless treasure chest of knowledge, truth, and wisdom? At the end of the day, you may never know if a psychic has really communicated with your energetic body (instead of inadvertently contacting negative beings in the lower-dimensional worlds), but you can *always* trust yourself.

This is not to say that all psychics are bad or are crooks. Some who have managed to successfully steer clear of alcohol and drugs, and live their lives in high states of spirituality, are indeed true energetic communicators. The rule of thumb for differentiating between them is, as always, to follow your own senses and intuition. If the medium's words and tone are laced with negativity, and he or she requires that you provide "offerings" for the "spirits" communicating with you, you can be sure that you are actually communicating with beings of the lower dimensions and negative ranges. The true denizens of the higher dimensions are so humble that they would never see themselves as being higher or better than you, nor would they care for your offerings or worship. The negative traits of pride and ego simply do not exist in the higher realms.

The Deception of Drugs

Drugs are a terrible poison to the spiritual evolution of humanity. Drugs can trick your mind into thinking that you have formed a true spiritual connection, but because this occurs in the absence of love and light, and without the presence of high-energetic frequencies, you actually become more susceptible as a target for negativity of all sorts.

Consciousness, in stark contrast to the term's academically endorsed definition, can be simply defined as one's level of spiritual evolution. The majority of people possess only a small fraction of consciousness, a portion that is capable of being increased through dedication and the training of the mind. Yes, you can actually increase your consciousness! It may be helpful to think of it this way: Another way of defining consciousness in its energetic sense is the linkage of one's energetic fuel to one's vibrational frequency. Therefore, in order to raise one's vibrational frequency, the amount of energetic fuel needs to be increased, as well. Contrary to current understanding, consciousness *can* be measured. The more one has, the more one has to divide between the physical mind and the energetic, thus enabling one to achieve conscious travel. When you consume drugs, that small portion of your consciousness that you use (about two-10ths) is split into two, with one-10th dedicated to the physiological mind and the other 10th dedicated to the energetic body. This actually explains why drug users are able to retain vivid memories of seemingly impressive so-called spiritual experiences while on drugs. In reality, however, these experiences are anything but impressive, for they are merely products of the individual's own powers of creation stimulated by the drugs.

The way drugs work in the energetic realm is no different from the way a con artist trained in the art of deception cons his victims. They trick the human mind into thinking that a spiritual connection has been achieved, when in actuality they have precisely the opposite effect, lowering the vibrational energetic frequencies of the user and automatically attracting him or her to the lower-dimensional existences. The hallucinations that drug users often experience, an effect attributed to the mind's power of creation, closely resemble human dreaming. Even from the very beginning of an individual's experimentation with drugs, the connection between the mind and the energetic body is completely destroyed, and hence his or her supposedly "spiritual" experiences are attained essentially by losing control.

As a user continues to use, the portion of consciousness allotted to the human mind shrinks to nil, while the energetic body redirects and increases to a whopping one-fifth, thus creating an imbalance that further induces hallucinations. With no consciousness residing within the human mind, the individual loses control over his or her thoughts, whose purpose is to serve as the steering wheel of the energetic body. This creates a disconnect between the heart and the mind, and transforms the user into an empty vessel overtaken by forces of negativity and low vibrational frequencies. Because traveling to the higher realms is impossible without love, these individuals end up much like cars on cruise control with no driver, and this is exactly why many addicts seem to do a complete "180" once they start using drugs. With all of this in mind, instead of seeking spiritual completeness or attunement with the energetic realms by falling into the deceptive trap of drugs, you can achieve this simply by concentrating on love and light!

Helpful Tips

How to differentiate between negative and positive beings

Differentiating between these two beings requires energetic discernment, as all communications in the universe are transmitted through vibrations and energies.

Positive beings of a higher nature will speak in a tone that instills the fundamentals of both love and light. When conversing with these beings, you will be able to clearly feel the love that resonates through the energetic realm and into the core of your very being.

Positive beings are free from the negativities of the lower worlds. Negative traits such as pride, arrogance, jealousy, and even desire simply do not exist in the higher realms. Both their personalities and the conversations they engage in are entirely free from these negative attributes.

If you embark on your travels with even the slightest negative emotion or thought burdening you, you are bound to encounter beings that are not of love and light, as the law governing all universes will not permit you to travel energetically to the higher realms unless your frequency is commensurately high. So, your state of mind and emotions upon traveling will have the biggest bearing on whether you meet positive beings or their negative counterparts.

Remember that your intuition is your greatest tool in the energetic world, so if your sense or intuition cannot say with certainty that the being you are encountering is positive, it is quite possible that you are dealing with a negative situation.

Indicators of the lower-dimensional existences

1. **Noises and voices.** Voices, chattering, even the sound of bells ringing can all serve as indicators of what dimension you are in. Because they have the ability to match

the basic rhythmic vibrations as those of the human energetic realm (or fourth-dimensional dream world), lower-dimensional beings can communicate directly with your energetic body. These communications are then transferred to your mind and registered as sounds formed through the process of hearing. On the other hand, higher beings of pure love and light that vibrate at much higher frequencies rarely, if ever, shift into the human energetic realm, unless they need to help someone.

2. **Color, light, and heaviness.** Colors, like sounds, are composed of rhythmic vibrations and hence serve as useful indicators of one's vibrational surroundings. The easiest way to comprehend the importance of colors in the energetic sense is to think of them as individual molecules, only energetically. Because one's unique vibrational color cannot be altered or hidden, it acts as a kind of "marker," enabling every being in the energetic realm to detect the level of spiritual evolution in others.

   In the energetic world, the color red tends to be associated with negativity, while gold indicates spiritual evolution and unity. Certain colors can be attributed to the natural interactions of thoughts or intentions with the primary energetic components of the ethereal worlds. Therefore, when you encounter portals in the form of tunnels or doorways during your travels, it is always best to use discernment, using this knowledge of color as a guide. As a general rule, blue and yellow serve as signals of the higher realms, while red and purple tend to designate the lower realms.

   As you gain knowledge and experience from your journeys, you will begin to notice that the higher and more positive the existence, the brighter and more vivid the colors you will encounter when traveling there. The rich, bright, vivid colors present in the higher realms are illustrative of the presence of life and energy in their purest states. If and when your travels take you to the lower dimensions, you will often find yourself enveloped in pitch darkness, as though you have been struck blind. This is due to the fact that discrete colors do not exist in the lowest dimensions, just as the forces of love and light have ceased to exist in these dark realms.

At times you may experience a feeling of heaviness and find it difficult or even impossible to move. Partial or complete paralysis means that the area you have travelled to holds such dense energies that even the energetic body, vibrating at a low frequency, cannot seem to resonate in tune with them. This dense energetic surrounding will latch onto your energetic body in an attempt to force it to vibrate at the same frequency as itself. What results is a distinct sensation of being magnetized or rooted to the spot by a strong invisible force, a phenomenon that is often referred to as *sleep paralysis*.

3. **Music.** In the movies and TV shows we watch, we often classify the voices of certain characters as "evil." This is not an accident. As you know, we are all composed of rhythmic vibrations, so when someone's overall vibrational frequency is especially low, even the energetic vibrations that they transfer as sound will come across as particularly strange, creepy, or evil. This is not to say that all people who have odd or menacing voices are vibrating at an extremely low frequency; sound functions as an accurate indicator of vibrational frequencies only when it is perceived and absorbed in the energetic sense, not the

physical. Because the energetic world functions on the basis of rhythmic unity, wherever you go you are bound to hear the energetic rhythm of that particular realm.

Exercise: Clearing Negative Thoughts

As a multidimensional traveler, it is extremely important that you clear your mind of all negative thoughts before going to sleep. There are numerous ways of going about this, but the most effective one in terms of sharpening your ability to travel is through meditation with the help of affirmation.

✳ While standing, take as many deep breaths as necessary to make you feel entirely relaxed.

✳ Each time you exhale, envision your negative thoughts slowly dissipating out into nothingness. Feel yourself becoming one with the force of love as you fend off every negative thought that comes your way.

✳ Envision yourself among the many universes, in the midst of stars and constellations and nothing else. It is just you in the galaxy.

✳ Now, repeat in your mind: *I cannot succumb to the forces of negativity. My life on earth is simply temporary, as I see the big picture. I know I am one with the force of love.*

✳ As you continue repeating this, imagine yourself being entirely enveloped by a whirlwind of vibrant, golden-yellow energy.

6

The Future

The little girl rested her warm forehead against the cold, misty window and stared out into the thick blanket of white snow that covered her family's farm. The icy breeze from outside had begun to send a chilling sensation through her head, but as it traveled south down her spine with its creeping force, she could not have cared less about it. Her thoughts were entirely somewhere else.

"Ohayo gozaimasu, *Michiko-Chan!*" Outside across the frozen expanse of greenery she spotted her neighbor, an elderly woman who still bore the lively spirit of a 20-year-old girl. Gleefully smiling and waving her hand excitedly in the brisk air, obasan *tried to get Michiko's attention, but Michiko did not reply, instead feeling a heavy ache of sorrow course through her heart. Stormy droplets of tears joined together and cascaded down from her almond-shaped eyes and across her ashen face in flowing ripples, as she yet again found herself drowning in thoughts of the future. Within the neglected depths of her mind, the same frightful future she had seen the night before was replaying itself.*

While asleep the previous night, Michiko had dreamt of her village being horrifically destroyed by an unforeseen earthquake. She witnessed obasan *attempting to flee the violently spreading water, but for what seemed to be like the very first time,* obasan's *old-age stamina had failed her and she was left to drown in the deluge within minutes of attempting to flee. Upon hearing of her nightmare, Michiko's concerned mother turned to Japanese tradition to prevent her despondent daughter from speaking of it. But the fresh wasabi she had grated and smeared across Michiko's lips was only a partial cure; it could not prevent Michiko from thinking about it over and over again.*

"It was only a nightmare," Michiko's mother's words entered through one ear and flew out the other, while her heart and mind remained stubbornly certain that what she had seen the previous night dwelled far beyond the depths of any dream. The nightmare continued to replay itself in clips through her mind, as if a movie were being projected through the window and out into the thick, icy haze.

When lunch hour came, Michiko darted outside across the snowdrifts in her bare feet. With a minkan *(orange) in tow as her offering, she was heading to pay homage to the forest spirits. Within her was the desperate hope that by doing so, she would divert the unfortunate future she had envisioned the night before from occurring.*

Far beyond Michiko's nightmare lay a horror of a different kind—one where among all the chaos, honking cars, pollution, crumbling uneven roads amassed with trash, and general dissatisfaction, a country stricken ruthlessly by poverty could be made out. In a neglected corner of the packed street sat a young blind boy on a makeshift seat he had improvised from the tin lid of a trash can, creating music out of used beer bottles and soda cans underneath the blazing rays of the sun. His lips were painfully chapped and he was unbearably thirsty, for he had not been rewarded with a bottle of water that day, as his earnings had been too low. To make up for these losses, he would have to earn at least 20,000 rupiah, or two dollars, by the end of the day, or else face being beaten with a leather belt by his master, a brutal gang member in Jakarta. As the scorching heat dried his lips even further, he began to feel faint. His head began to spin uncontrollably, and as he desperately clutched on to it with one hand and continued to slap the empty glass bottle against the pavement with the other, it felt as though the life-force were being sucked out of him. But beyond this reality of physical earth, within the calming depths of his mind, were thoughts of the future, and it was only these thoughts that could bring him any glimmer of solace and comfort. He couldn't understand why, but thoughts of the future brought him happiness and peace. He wasn't aware of the past anymore, nor could he remember it, and he ran instinctively from the present. It was only the future on which he centered his thoughts and looked forward to with immeasurable anticipation.

Creating the Future

What exactly is it about the future that greatly entices all of us? Perhaps our dedicated focus on the coming future lies in knowing, deep within our hearts, that we will eventually break free from our physical encasements. Perhaps, buried within the hollows of our hearts is an understanding of the true energetic definition of what we call the future—the evolution of our energetics into a higher-dimensional world. *This* is the future that we look forward to, not the future that invitingly offers us the temporary pleasures of physical life.

People near and far remain oblivious to the everlasting existences of the energetic worlds. We are taught to simply focus on the pleasure we attain from the physical aspects of our temporary lives on earth, and as a result we find ourselves in a conditioned rat race, pondering the "missing piece" of our being. We seek the company of other people, search for the perfect lover, eat delicious foods, and strive for material wealth, all for the purpose of finding something, or really anything, to fill up the perceived gap within us. And even when it is finally filled, it remains veiled by human emotion, until the feeling of incompleteness creeps up behind us once again. This emptiness is one that can only be fulfilled through accepting and cherishing the knowledge of our true energetic nature.

We often regard our actions and the decisions we make as bearing the fruit of what we call the future. In a sense this is true, but our actions are also governed by our thoughts, and the two are, in fact, dependent on one another in the physical. Although our actions are always bound to contribute to the future, we must not forget that because our true nature is energetic, our thoughts also play a major role in the manifestation of the future.

By now you have most likely come to understand the importance of thoughts in relation to the energetic world. It is with this understanding in mind that you must realize that when you produce thoughts, you are actually manifesting occurrences or events within the energetic realm. Our thoughts, along with the courses of action in the universe, create energetic "imprints" of the probabilities of the coming future. It is due to these energetic probabilities that we can trust with confidence that the future is *not* set in stone and can be altered through the conjoining of thoughts with the necessary courses of action to create alternate motions in the energetic worlds.

When someone claims he can see the future, he is viewing this energetic probability in the context of the human energetic dimension, which you yourself can travel to by gaining expertise and furthering your own experiences in the multidimensional realms. It has become a sad reality that certain people, whose aim is to achieve control over others, have popularized the false idea that some possess a special and unique ability to glimpse into what is known as the Akashic Record—a record that exists in the etheric realm and contains the imprints of everything that was, is, and will be—the past, present, and future of all planets and beings. Although the Akashic Record, whose name stems from the word *akasha*, Sanskrit for "etheric," was originally recorded in sacred texts for the purpose of revealing the true nature of the energetic realms, these people are completely misinterpreting it for their own ends. It is indeed true that every one of us possesses the ability to travel multidimensionally to the human energetic realm and to our own destined energetic probabilities through the elevation of our vibrational frequencies, but it is against the laws of the universe itself for any one being to possess the knowledge of an entire history that could put others in grave danger. You must become aware of your innate ability to freely travel to the energetic probable future, and never deem others as "special" for possessing this ability, despite their claims to the contrary, as this ability exists within each and every one of us, whether we know it or not.

The Museum

"This, here, is a bowl that was used to turn corn into powder," my archaeology professor monotonously recited, as a bright red dot from her laser pointer fell upon the projection screen in front of the dimly lit university lecture hall. A few heads turned from their laptops and cell phones to look at a slide of an Incan remain found in Peru, while the chair next to me let out a squeaky creak. A girl leaned back in it, her legs pushing off of the chair in front of her, and began to tap away on her cell phone, the glare of its screen reflecting in her eyes. As our professor continued her explanation of the artifact, a few students sitting toward the front of the lecture hall began to hastily chop away on the keyboards of their laptops.

I have always loved museums, with their vast, intricate collections of archaeological treasures. They and their contents have captivated me for as long as I can remember. But whenever it comes to the tiny, eight-point-font descriptions on the other side of the glass, I have never bothered to read them—not so much because I would have to squint my already-strained eyes, but because I know that they would only add to my growing frustration.

I had taken plenty of other archeology classes before, and every one of them would always fall out in a similar manner—me, sitting toward the back of the lecture hall, and cringing with disappointment at each one of the descriptions of the excavated items shown on the professor's slides. That day, when I had finally had enough of listening to human misconceptions and perceptions, my right hand shot up in the air. In the lecture hall full of caffeinated, sleep-deprived students, the professor's gaze soon picked me out. "Professor, how do you know this, considering that none of us was alive during that time? Was there a kernel of corn found in it that made you come to the conclusion that it indeed had been used to grind corn in the past?" The professor instantly pursed her thin lips and shifted her gaze at the screen in a sense of alarmed disorientation, before turning back to me. Everyone was now gaping at the front of the lecture hall, all previous clicking of keyboards suddenly silenced, to catch a glimpse of the professor's reaction.

"It is what the archaeologists have deemed it to be. Why? Do you have any other thoughts that you'd like to share?"

I hesitated for a split second before explaining, "Well, if we only see the sacred objects of the past for what they physically look like and connect them to our present time, considering that it is a different time, how will we be able to discover the truth behind the use of any one object? For example, say a catastrophe occurs tomorrow, and far in the future someone just like us comes across a Coke bottle and assumes that we had a god by the name of 'Coke.'" The girl sitting next to me giggled quietly as the professor stared questioningly at me from the front of the auditorium. "I mean, they could perceive this bottle to be anything, really! But that's wrong, then, isn't it? Doesn't connecting and molding such objects to our own beliefs, rather than welcoming free thoughts about them, only lead to misconceptions about the actual use of the object?" I peered at the professor from a distance, while the rest of the class watched carefully, turning their heads from one of us to the other as if they were watching a Ping-Pong match. The auditorium soon erupted in chuckles and giggles, and the professor moved on to the next slide, pretending to have not heard my comment.

You can take it from me. It's a difficult task to constantly try to break free from society's beliefs and perceptions. Even when you manage to do so, you'll most often find yourself being scorned. But the more you practice seeing things through your own unique perspective, rather than that of the collective, the more you will attune yourself to the right vibrations that are free from society's expectations and pressures. Imagine how much extraordinary knowledge and previously ignored truths would arise for the future generations if only 2 percent of earth's current population began to question the knowledge that we are constantly being indoctrinated with and spoon-fed. But to be able to do so will, first and foremost, require us to digest and ultimately discard this knowledge that we have ignorantly and indiscriminately consumed, because for however long this collective thinking continues to exist within us, we will not be able to truly experience the limitless abilities of our own minds.

The Museum of the Future

By my senior year of college, I had already ventured on plenty of journeys to the energetic probabilities of the future. Some of them had occurred naturally; others I had executed through the sheer power of will. I had learned and thoroughly come to terms with the importance of trusting the universe and putting the thought out for it to guide me toward my fated lessons while in my energetic body.

"I wish to travel to future earth, one that I could learn and attain wisdom from," I recited while in my energetic body one evening, after successfully releasing myself from my physical constraints. I was transported to a long, narrow hallway intersected by many corridors. Having no idea of where I was, I shuffled down the seemingly endless passageway until I turned into one of its inviting corridors, only to find an immaculately clean and shiny glass case; inside was an incredibly realistic-looking sculpted figure of a Neanderthal grasping a stone club. A description stuck on an upper corner of the glass case read clearly: "Our ancestors from the past were long aware of the multidimensional realities of our world. Around the world, tunnels have been found with depictions of other worldly beings."

I felt elation and pleasant surprise resonate throughout my being, as my eyes reached the last word of the description. With great curiosity, I continued on to the neighboring corridor to find yet another glass case, this one containing a bronze-colored stone plate portraying some sort of being. As I moved excitedly from corridor to corridor, it became clear to me that this "museum of the future" had been created based on the knowledge of our multidimensional nature instead of egoistic human history—a history entirely devoid of the truth that humans are only a small part of this infinite universe. This museum was dedicated to the multidimensional nature of all humans, and as I stood there, taking in all the wonderful truth that it openly offered, I was overwhelmed with astonished joy that such a museum would someday exist.

I soon came across a separate wing that was dedicated entirely to the musical instruments used by our ancestors to raise their vibrational frequencies through music. Not a single one of these objects was degraded as being "agricultural" or as something that was used by "less-civilized" people, as they commonly are today; instead, the descriptions emphasized each object's relation to the energetic and multidimensional world. One object in particular drew my attention with a magnetizing attraction. I had seen it before in a museum in Peru—a pendant-sized flute that, rather than being simply an ornament or a piece of jewelry, as current thinking would have us believe, was in actuality used as a form of protection whenever someone would find him- or herself in a lower-vibrational frequency. Attached to a sturdy thread that was tied around the neck, the instrument could be played by the wearer to uplift his or her vibrational frequency and prevent the energetic body from being forced into a lower dimension.

In the midst of countless stultifying discussions about excavated bowls used for grinding corn and a discouraging sense of dim despair that future generations would continue to live blind to the multidimensional realities, the universe had given me a spark of hope for humanity's progress in its spiritual evolution. What had at first seemed to be a simple and uneventful journey had ignited an undeniable flame of determination within me to continue the work of revealing the true energetic identity of our universe.

Kyoahin

"Did you know that if you open your palms and connect the lines you see on them, you'll be able to see the shape of your future husband's face?" my eldest sister exclaimed with enthusiasm as she lifted her opened palms in an excited attempt to introduce her "future husband" to my family. It was Chinese New Year and all of us were gathered in the living room that had been meticulously decorated by my mother at least a month in advance, playing all sorts of games and sharing our superstitions about the future. "Well, this year is the year of the dragon, so you're lucky, Khartika. And as for you...," my mother gleefully shifted her sympathetic gaze from me to my little sister so that she could repeat her usual yearly sentiments. Before she could manage to do so, however, my sister interrupted her, "Yeah, yeah. I know: The sheep gets burned or eaten by the dragon, so let me guess: I'll have bad luck again this year? It's inevitable, isn't it? The sheep will always get eaten by everyone except for the pigs, chickens, and rabbits. I *hate* being a sheep!" She rolled her eyes visibly, perhaps so that my mother would finally take a break from opining on her dreadful future. Instead, a sheer look of horror overcame our mother's ashen face as she lifted her forefinger to her lips and snapped, "Shh! Blasphemy! You don't want to give yourself worse luck this year, do you?"

"Enough, enough. I see you, granddaughters, leaving some rice leftovers on your plates. If you finish all the grains of rice on your plate, you'll have a handsome spouse with a face as smooth as silk. But if you don't, then don't expect any pretty boys!" my grandmother interrupted, shooing us back to the table with yet another illogical superstition that has been passed down by Chinese families over countless generations.

My older sister, who seemed to have already forgotten about the husband she had seen in her palms, darted ahead of us toward the dining table to finish not only her rice, but also to clear everyone else's plates. She had dedicated her entire life to chasing boys. "Khartika, you're not finishing yours yourself?" my grandma asked me in a jokingly scolding yet genuinely inquisitive tone.

"No, Grandma. I concern myself with the world and my career, not silly love life," I explained softly with a cheeky grin, to which she simply responded with an audible sigh.

Later that evening, tired from the day's celebrations, I stepped into my room and immediately adjusted my air conditioner to its lowest possible setting before preparing to go to bed. I knew that the colder my room was, the easier it would be for me to travel undisturbed. In what initially seemed to be a typical night of multidimensional traveling, I found myself roaming excitedly around the energetic world, with the general idea of looking into the future. Previously, it had been the course of the universe itself that had led me to traveling to the future, but this time I was determined to use the power of self-will in the energetic existence to do so.

I could not tell how far into the future I had traveled, for determining time is a particularly tricky task when it comes to the energetic world; you cannot simply ask the universe to pull your energetic body to a specific time. I found myself surrounded by beautiful, cascading mountains in every direction. Deep within the lush greenery of the sloping hills, I could make out a futuristic-looking yet humble wooden house shaped like a parallelogram. I walked inquisitively through the house's front door and found myself unintentionally interrupting a family dinner. At the round table sat a teenage girl, her jet-black hair thrown neatly over her shoulders, along with her parents. They were all dressed similarly, in loose pants and buttoned shirts that appeared to be made out of hemp. As they gazed at me in conscious curiosity, their mouths open in surprise, I soon realized that I, too, had been standing there with the same expression on my face. Quickly composing myself, I asked softly, "You can see me? Is this earth?" The family nodded in unison, with their mouths still hanging open. "How are you able to see me? I thought that all of you were in your physical bodies. My name is Khartika. I come from the earth, year 2009."

Upon hearing my words, the girl shifted slightly in her wooden chair and turned to face me. "We are in our physical. Since you come from the year 2009, a long time back, you must not know that by now we have elevated our spiritual understanding and are able to interact with those in the energetic world. Forgive us, I...I mean we..." The girl paused briefly to nod at both of her parents. "We have just never seen anyone from the past before. As a spiritual energetic projection."

The mother stood up suddenly and quickly moved toward the back of the dining room to find another wooden chair for me to sit on. As she kindly gestured for me to take a seat at the table, I could not help but chuckle to myself at the irony of being offered a physical chair while in my energetic body. I placed my hand over my mouth and pretended to cough in attempt to mask my laughter, but was soon distracted by a poem framed on the white wall of the family's abode. It read: *Goodbye, my love. My time has come. It is time for me to leave. I do not want tears or sadness, but a love that is everlasting....*

"It is called a Kyoahin spring poem," the slender mother explained, with kindness emanating through her hazel eyes and a grin cast across her face. "It is a famous poem that speaks of spirituality and love. Kyoahin is the name of the genre of poetry that conveys our energetic nature. Kyoahin is also a movement, a movement to reconnect in the energetic worlds and to return to the revival of our energetic realities. I guess it would be easier for you to think of it as a spiritual movement."

"I don't understand. I have seen the future before. I saw the energetic probabilities of it. I saw developed nations, such as America, transformed into rigidly controlled countries, with police patrols everywhere and freedom taken away from the people. I saw a draconian ruling class, one that entirely suppresses the freedom of the mind. I saw many disasters. Did any of this occur? Is this an alternate future? What year is this?"

As I stared at the poem in confusion, I heard the father respond from the table, "There is no year; we don't count by years anymore. We live our lives day to day on the basis of spirituality. We try to put away the spiritual aspects of time. What you saw wasn't altered. It happened in our past, but we are now far into the future from that period."

I began to look around myself in search of any futuristic gadgets, eager to make more sense of the man's words, but my efforts were soon interrupted by the mother. "We don't entertain ourselves with the entertainment that people in your period once did. We focus entirely and wholeheartedly on spirituality now; we have gone back to our origin. The entertainment of the past served only as a distraction from the spiritual path and the true knowledge of our temporary physical existence. Currently, we focus on spirituality. What you once knew as consumerism has ceased to exist. We grow our own foods and exchange our own goods with one another through kind gestures of giving and receiving. We do not have what you once termed as *shopping*; such was only yet another obstacle for blinding humanity to its purpose and true nature. We now focus on spiritual attainment." She briefly excused herself from the table, before returning with a framed picture that she handed to the father.

He gently placed the framed picture on the smooth table in front of us and explained, "Look. This is the map of our earth today. As you can tell, many countries have physically shifted their positions since your time. See?" The father intently placed his forefinger on the map and looked at me to make sure I was following along. "Cambodia, Laos, Thailand, and two other countries formed a single alliance and formed into one. That is where you are now, right here. We are called Iona now. The land that you are standing on was previously known as West Sulawesi." He hovered his finger in a circle above a particular region of the map. I looked at it blankly without even the slightest sense of recognition. It looked nothing like the earth we know now: Each continent looked as though it had either been altered in shape or moved entirely.

"Is it just you who live in this vast area of space? Or do the people of your current period prefer to live in solitude like this?" I asked with my usual curiosity, as my eyes roamed the map still in front of me.

"There are a few reasons, but the primary reason is that the people of our time recognize that we do not need such large spaces to live in. We focus on living humbly, if your question lies in connection to the size of our house. Currently, earth is much less populated than it was during your time. In fact, we learned that earth was most densely populated during the clouded period that you live in."

"Clouded period?" I asked, as I looked up with my brow crinkled in confusion.

"That's what we call your time. After the clouded period comes the period of darkness. It is known as 'clouded' because during this time most people's minds were clouded, blinding them from seeing the truth of the importance of spirituality, and instead—inviting them to dedicate their efforts, time, and mind to the attainment of material wealth." The father's light green eyes overflowed with a dim grayness, as I took in his words.

"I know. I feel the same way. I am presently thinking of writing a book, a book to help rekindle the truth of the energetic worlds and the true energetic nature of humans. Maybe, perhaps, the clouded period will only pertain to the majority of people living during my period, but not all. Because I know strongly within me that a handful, though not many, will begin to contribute in the reigniting of the energetic realities and help maintain energetic harmony during my time," I said with a smile full of hope, which the father kindly reciprocated.

"You'd be happy to know that even the schools, in which everyone is entitled and encouraged to follow their passions, teach about the energetic realities and the energetic worlds. Also, did you know," the girl beamed excitedly as she plaited her hair into a neat braid, "that we can now pick our birth time, and that the last name we receive depends on it? I was born in the afternoon, so my last name is Guna."

I was blown away at these words, and replied in astonishment, "This is amazing. It is just so hard to believe, coming from my time, that a population could reside within their physical bodes, live on physical earth, yet maintain happiness and peace. Does crime really not exist?" I apologized for my apparent negativity before they could answer my question, explaining that our period was so drastically different from theirs in every possible way.

"Crimes cannot exist solely due to human nature, without external inputs or influences. Even if a crime does occur, the criminal himself knows immediately that there is no place for such evil doing in our society. Everyone here vibrates in a high frequency. Therefore, if one person vibrates in a low frequency, he must either leave to maintain his own frequency, or he must elevate it so that he can vibrate in the frequency of others," the girl explained.

I thanked the kind family for evoking a hope for humanity within me and for showing me a future that gives me happiness in the present. I asked politely if I could take another look at the Kyoahin poem, at which the mother bounced up from her seat and made her way over to the wall with the poem. As she handed over the simple wooden frame that rested in both of her hands, I explained in gratitude that I would try to memorize it so that I could remember it upon returning to my physical body. The family chuckled good-naturedly, and the mother responded softly, "You cannot possibly remember all 20 lines! Just focus on five or six." I then asked if I could know the name of the author, to which she replied, "We do not know. That is the entire essence of the Kyoahin poem. Its nature is that of spirituality and its writing is for the purpose of reviving spirituality. The writers of Kyoahin do not provide their names to gain recognition or fame. It is believed that by writing a Kyoahin poem, you are elevating your energetic frequency through assisting in the elevation of the frequencies of others. The more you give, the more you receive."

"Maybe I'll try to remember as much as the universe will allow me to. I'll write it in the book and then anyone can finish it for me," I responded.

I recited the poem about 10 times to myself. As my eyes skimmed up and down the framed poem, I felt my chest immerse in a joyous light sensation of tranquility—a feeling of unity, oneness, with the universe. Suddenly, the discordant ringing of my cell phone cut through my mind, and I hurriedly thanked the family for their hospitality as I felt myself being rapidly pulled into space. I drifted farther and farther away until I was at a distance where I could clearly make out the outline of the country I had just visited, as though I were looking at my very own map. I was shocked to notice a group of iceberg-like formations floating around the region. But who am I to question the energetic probabilities of the future? Am I not but a caveman of the past?

The moment I found myself back in my physical body, I quickly swiped open my phone, my fingers running restlessly over its glowing screen. I immediately wrote down the experiences of my journey, but was rather disappointed at myself for being able to remember such few sentences of the beautiful poem. Releasing an exasperated sigh at my faulty memory, I dialed Mary's number and, after reciting whatever I could remember of the poem to her, I asked her what she thought of it. "Hmmm...it's like... I know! Romeo and Juliet! Right before—before he drank the poison." Irritated, I hung up abruptly, later explaining to Mary that my battery had died. I couldn't let a trite analogy of romance ruin the serene spirituality of the Kyoahin spring poem.

It is likely that the part of the poem that I was able to recollect and share with you here does not do justice to its beauty, as it was only the first few lines of a perfect whole. Poetry itself is meant to be felt and experienced in its entirety, but I can tell you this much: It was quite easily the most beautiful poem I have ever read.

Helpful Tips

Breaking free of the fourth-dimensional realm

Although we are free to travel to any of the dimensional realms we wish by adjusting our overall vibrational frequency, we most often find ourselves being pulled automatically into the fourth-dimensional realm. When any being or group of beings evolves spiritually, they will be energetically elevated to a dimensional existence of a higher nature. This same natural energetic law applies to humans: As more humans attain elevated levels of consciousness, the likelihood of their energetic bodies being permanently elevated to a higher-dimensional existence drastically increases.

The human race is unlike any other that exists intradimensionally. We exist as a hodgepodge of beings, all possessing differing levels of consciousness on the singular planet. Many of you possess higher levels of consciousness than others, and you should know who you are. Whatever your physical surroundings and circumstances may be, you must focus on individually raising your level of consciousness and maintaining a consistently high overall vibrational frequency so that your energetic body can travel to and exist in the higher- dimensional energetic realms, instead of remaining bound to the fourth-dimensional realm.

Directing your consciousness to places you wish to visit

As a multidimensional traveler who vibrates at a high frequency, at times you will find yourself being pulled into strange realms and existences. It is important that you do not allow fear to creep into your mind during these moments, and instead observe and analyze the situation with a higher understanding that everything you experience is a lesson for you to learn, a task for you to master. As your energetic abilities grow with the many lessons the universe brings to your feet, your ability to directly focus your consciousness on traveling to a location of your choice will also increase. Raising that fraction of consciousness within your energetic being will increase your ability to retain the information you receive while traveling in the energetic realms. The same also applies when you direct your travels to particular places within the energetic world, as an increase in your overall level of consciousness is accompanied by a larger fraction of consciousness in your energetic body, providing you with enough control to direct a small portion of your consciousness to the desired location of your own choosing.

When you possess enough consciousness within your energetic body, you will be able to "steer" yourself to whatever planetary realm you wish to visit in the energetic world. A small fraction of consciousness existing in your energetic body will be directed to the realm itself and serve as an automated temporary connection that will enable you to align your vibrational frequency to the rhythm of the planet. For precisely this reason, it is crucial that you use discernment when choosing a planet to travel to, as you most definitely will not have a pleasant experience on planets whose frequencies are much lower than your own. As we are admittedly still fairly limited in our knowledge of the various energetic existences, instead of directing yourself to any particular planet, it would be wiser for you to instead relay your general wish to travel to the higher realms of love and light, out into the universe. When it comes to our consciousness and even our energetic bodies, we must put aside our human understanding and physical biases, as consciousness in the energetic realms encompasses and exceeds all possible physicalities.

Directing your consciousness to future energetic probabilities

Of the many mistakes that people make when attempting to travel to the future, the most common is commanding their mind to do so before they fall asleep. Doing this actually creates a self-induced future in the human energetic or dream realm, due to the automated sequencing—uncontrollable thoughts—that manifest events in this realm. People who do this are actually seeing the energetic projections of their own auto-mated states perpetrated by the mind and derived from their subconscious thoughts. They have no idea that by doing this they are simply living within the energetic realms of their own imaginary, self-created ener-getic world—or, in other words, dreaming. Therefore, you must direct your thoughts to traveling to the future only when you are fully aware and conscious in your energetic body and after you have entirely sepa-rated from the physical. It is only after you achieve a state of awareness and alertness in your energetic body that you will be able to instruct your mind to venture wherever you desire. This will ensure that you have absolute control over your energetic body.

Attaining complete control of the energetic body is incredibly differ-ent from what we refer to as "dreaming." This distinction is based on the immense importance of taming your mind on a daily basis. It is through taming your mind that you will be able to prevent it from continuously imposing subconscious manifestations on your energetic body in the dream realm, and gain enough consciousness to direct your energetic body to the boundless future.

Despite what any sci-fi novel or Hollywood film may suggest, when in comes to traveling energetically to the future, you cannot simply direct yourself to any particular time or place. Instead, you must know that whatever future you find yourself in is the one that has been fated and allowed by the universe. Remember that the concept of time ex-ists only in our physical world; the future itself is entirely free of such constraints.

Manifesting the future and "luck"

We tend to concentrate on the future without realizing that by doing so, we are actually constraining ourselves by the humanly created barrier of time, a concept that does not exist in the universe. In thinking about the future, many people erroneously assume that the events of the future depend on them, when in reality the future is a manifestation of the natural occurrences of the universe. Therefore, instead of relying on the future, we must focus on heightening and maintaining our vibrational frequencies, which in turn will manifest a positive future. *This* is how we make our own "luck."

7

Attacks

Lotuses in a Dirty Pond

As a young girl, I spent most of my afternoons at my grandparents' home in South Jakarta. There was simply no other place that I'd rather be. Though they could have easily acquired a larger and much more luxurious house, their home resonated strongly with their values of modesty and simplicity. They had always provided me with the utmost love and, in addition to that, an afternoon retreat that was both humble and cozy—just like them!

My grandfather was the most disciplined man I've ever met. His days were planned with such remarkable diligence that I often wondered if there was ever a day when he had mistakenly slept in. At exactly 2:30 every afternoon Grandpa would pluck the ripest guava from one of the many lush trees in his carefully tended yard and cut the sweetest parts for me to have, leaving the less-desirable parts of the fruit for himself and Grandma. Even at such a young age, I was incredibly blessed to have understood and experienced the unconditional love at the root of my grandparent's every action.

But perhaps my favorite time of the day came at 3 p.m., when my grandfather would make his way to his small pond, which rested sheltered under the shade of the scattered emerald-green guava trees, to feed his beloved koi fish. Though my memory often proves to be faulty, I can recall one such day with absolute vividness: my grandfather pouring buckets of food into the gleaming water for the fish that danced only inches below the surface, and an 8-year-old me standing giddily in one of the corners of his award-winning pond, directly under the scorching rays of the tropical sun. I giggled loudly as I watched the colorful fish open their mouths shamelessly at my grandfather, a man who had never let them down when it came to lunchtime.

"Grandpa," I peered at my grandfather through round glasses that were far too big for my small face. "The pond is so clean and beautiful, and the fishes are so pretty! And the lotuses around it just make it even more beautiful!" I shifted my focus back to the greedy fish, but through the corner of my eyes could see my grandfather beaming with pride.

A smile smoothed his sun-crested face, and he said to me in his usual tone that radiated both calmness and wisdom, "It's only the surface that you're looking at, my granddaughter. Underneath and all around, it requires a lot of cleaning. It's dirty, in fact. But it is in such an environment that lotuses thrive, grow, and blossom."

Of course I believed my grandfather, but being the unsatisfied seeker of knowledge that I am, I set out to test his claim for myself. I shuffled curiously toward the edge of the pond, crouched down to the ground, and leaned over to place my hand deep within the water, all the while gleaming amusingly at my reflection blending in with the vibrant orange, white, and red of grandfather's koi fish. As I impatiently pushed my arm further and further into the pond, I could not detect even a single trace of dirt—until, within a flash of what seemed to be a second, I found myself inside of it, drenched from head to toe.

I clambered up to catch my balance, my small feet continuously slipping on the blanket of debris that conquered the bottom of the pond. Grandfather had been right, and I recall him roaring with laughter as he stretched out his strong hand and I desperately grasped onto a few poor lotuses for support. "Look! Even your white shirt has turned grayish-brown now! I did tell you that it's only the surface that is clean and pretty, and that the base of the pond is so dirty. It's the rainy season now; what do you expect, my granddaughter?" he exclaimed joyously, as I shot him a teasing pout. He crinkled his eyebrows in thought as I wandered back into the sun in attempt to dry myself, and I saw a toothy grin take over his face. "Hmm... though I wonder if you had not fallen in by accident, whether you would've gone in to see for yourself, my dear granddaughter. Does this incident make you think differently now of the pond being beautiful and spotless?" I kept quiet for a second or two after hearing his question, staring up into the sky and extending my arms out to my sides while I thought of a clever answer to reply with.

"Well, Grandpa, I don't think I would have gone into the pond myself. But you did say it's the rainy season, and during the rainy season there are tadpoles everywhere. So sooner or later, I would've spotted the tadpoles and tried to chase them, and then probably fallen into the pond and discovered how your beautiful pond actually is...a pond full of dirt!"

Grandpa let out a whole-hearted chuckle upon hearing my response and shook his head playfully at me, as I waved my hands in the air in attempt to dry faster. "Well, say that it's not the rainy season. What if you weren't able to spot the tadpoles?" he asked me.

Taking no time to think of my response, I replied to him cheekily, "But grandfather, I know I would've spotted them one way or another because either you or grandma would've pointed them out to me! Both of you always do! Grandpa, you lost!" I giggled. "One point for me and zero for you—too bad!" Grandpa nodded to me as though in surrender, and we slowly walked hand-in-hand back to the house to read a book together.

In this universe of limitless possibilities, where the true essence of our being lies in the sphere of the energetic world (which most people have yet to discover, let alone comprehend and accept), it is only the "tadpoles" that I can point out for you to see. It is up to you whether your own determination and curious mind will lead you beyond the surface of our world to discover the truth, or whether, upon spotting a tadpole, you will accidentally trip and find yourself entirely immersed inside the whirlwind of truth.

This story illustrates how we can see beyond the deceptive appearance of the physical world—the surface of the pond—and often trip and fall into this whirlwind of truth as we seek knowledge. Often, part of the result of seeking knowledge is discovering the ugly side of life on earth and that appearances are deceiving—which brings me to the topic of this chapter. Most people are unaware of energetic attacks because they are overly focused on their physical life here on earth. For as long as they refuse to seek the truth and look beyond the surface, they will continue being targeted by such attacks without even knowing it.

The Story

It was at the very beginning of my travels to the higher realms that I found myself continuously encountering numerous obstacles. At the time, I hadn't even the slightest idea what was happening to me when I would suddenly find myself in the lower-dimensional realities against my will or attacked by unseen electromagnetic forces. I've since come to learn through my experiences in the energetic realms that all attacks on the energetic body are electromagnetic in nature.

I used to think that it was solely humanity's unwillingness to see beyond the superficialities of our world and partake in spiritual endeavors, that placed our race far behind all others of the universe in the journey toward spiritual evolution. But I quickly came to learn the error in this perception. I would soon discover that the reality was much harsher than I had ever imagined, that it wasn't humanity's unwillingness to look beyond the surface of our universe that has kept our abilities dormant, but instead the vile individuals of our world who have managed to discreetly acquire and use weapons to keep our minds imprisoned and blinded from the truth. These weapons and brutal technologies have been specifically designed to suppress human consciousness, overstimulate the subconscious, and inflict pain on the human energetic. Unfortunately, due to this sad reality, the multidimensional traveling of today is much more challenging and different than it was even half a century ago.

These dark individuals have foreseen the future energetic probabilities of our planet, when we finally begin to awaken our dormant abilities and make discoveries of our energetic abilities. They have realized that it is precisely during this moment in human history that we will begin to shut the door on the illusion-driven physical world and unlock the door that leads us to our energetic origins. In order to continue exerting their control over the human race, these dark beings have resorted to doing everything in their power to prevent the human race from coming to this realization, not only through maintaining low-vibrational frequencies among humans (to suppress our abilities to travel), but also by attacking us in every way possible to prevent us from traveling further.

These dark individuals have targeted our vibrational frequencies and gone about lowering them in myriad ways. Although they have inflicted these brutal attacks on countless numbers of people, their victims can be divided into two major categories: 1) those who are satisfied with their physicality and remain oblivious to the truth, and 2) those who have just begun to discover their energetic nature. In the case of the former, the perpetrators of these attacks have sought to maintain the low-vibrational frequencies of these individuals through simply stimulating their subconscious with false information. This is often achieved through the repeated use of social propaganda, which is usually centered on the trendy use of drugs and the fashionable, excessive consumption of alcohol. The latter category of victims (who actually pose the greater threat to these dark beings) are attacked in numerous ways with the primary aim of throwing them off their paths to discovery and spiritual evolution. This second category of people experience both conscious and unconscious electromagnetic attacks, thus preventing them from achieving conscious multidimensional travels.

Hooded Beings: Part 1

After being flung into the energetic world almost without warning, I was finally starting to get the hang of traveling energetically. Most importantly, I had acquired a solid, general understanding of the keys to energetic travel, those derived from feelings of love and light. As summer wrapped itself up and prepared for its annual hibernation, I booked a last-minute flight to Chicago, where Marcia, a dear friend of mine, was preparing to begin her graduate studies. After a late night of catching up over takeout, I foolishly dozed off into a deep sleep feeling rather irritated and upset, after receiving a confrontational text message from my father. Rather than identifying these feelings with caution, as any adept traveler would, I chose to suppress them and instead dismissed them as mere fragments of irrelevant thought into the depths of my mind.

Not long after, I found myself submerged within a faint murmur of indistinguishable sounds and a burning sensation spreading ruthlessly in my chest. I frantically opened my eyes, thinking that I was still in my physical body, but as I struggled to turn my heavy head to the side, I realized that I was enveloped in pitch darkness and in an entirely different dimension. With a loud whoosh, I was instantaneously transported by an unidentifiable force. My entire being felt stuck, and as I urged my energetic body to defend itself from the straining magnetization, to no avail, I heard the chanting of "Ohm...ohm...ohm," at first just murmurs, and then increasing inexorably in volume. With every chant drilling into my mind, I felt the burning sensation escalate and move from my heart to my solar plexus. I gazed up desperately, my mind flooded with frantic terror, only to find myself surrounded by dark, hooded beings. They appeared as clusters of black shadows, their faces shielded from sight by their hoods, and though I could not even spot their mouths, I assumed that they were the ones chanting. As soon as I noticed their presence, I felt every inch of my being burn with a scorching intensity, as though I were being jabbed with a thousand flaming needles.

I tried to focus on feelings of pure love, replaying episodes of the happiest times of my existence on earth within my mind, while desperately begging my physical mind to wake up. I woke up to a deeply concerned Marcia hovering over me and repeatedly asking me if I was all right, her hands grasping onto my shoulders and shaking me with all her might. Thankfully, I had learned early on in my travels to inform whomever I was staying with to do whatever they had to, whether shaking me or splashing cold water on my face, in order to wake me up from my "nightmares" as soon as they heard me sleep talking. "Thanks, Marcia," I sighed heavily with relief and managed to force a faint smile across my face in gratitude. "What did I say, specifically?" I asked her, as I reached over the side of the futon I had been sleeping on to grab the thin quilt that had probably slipped off amid my frenzy.

Marcia disappeared into the kitchen, and a minute later returned with a glass of water and an answer to my question: "You sounded like you were crying. I tried shaking you several times, but you wouldn't wake up, so I slapped your face," she explained apologetically, as she let out a wide yawn. I reached over with my right hand to grab the glass of water she had brought for me, while placing my left hand gently over my solar plexus. As I laid wide awake on the futon in Marcia's living room, I felt that same burning sensation last for about 30 minutes until I finally drifted back into sleep. I didn't encounter the dark beings again that night, but it wouldn't be the last time I saw them.

Hooded Beings: Part 2

On a breezy autumn morning in late November, I found myself in the ancient city of Jogjakarta, an acclaimed gem of Indonesia. I was there with my mother, who had asked me to celebrate her 46th birthday there with her. In a city full of myriads ancient ruins, ranging from the famed majestic Borobudur to the mysterious Prambanan, my mother was dying to taste Jogjakarta's most celebrated and authentic dish, *gudeg*, a traditional rice dish consisting of jackfruits, eggs, chili, and chicken, the epitome of Indonesian cuisine. Being quite the obedient daughter, I agreed and accompanied her there for her culinary adventure.

We spent most of our first day in Jogjakarta eating my mother's long-awaited *gudeg* and venturing through the most popular ancient site in Indonesia, the Borobudur. Despite its towering stone platforms and grandiose *stupas*, it was only at night, when we returned to our hotel room and I drifted into sleep, that things began to get interesting.

"Ohm...ohm...ohm" the now familiar chant echoed through my mind in the same low and disturbingly unpleasant tone as before. With it, I felt my energetic self being pulled out of my physical body, and found myself in the same dark dimension with the hooded beings. Following my first encounter with these dark beings, I had dedicated many hours to researching and properly identifying the chants that I had heard. I learned quickly that the wise ancient peoples of the East had used *ohm*, a sacred syllable in both Buddhist and Hindu chants, for centuries for the purpose of uplifting their vibrational frequencies. But as the academic articles and blogs flooded the search engine page on my computer screen, a central truth that had been submerged deep within them revealed itself to me: Just as chanting the syllable *ohm* could raise one's vibrational frequency, when used the right way and in the correct frequency to promote positivity, it could also have the opposite effect, if chanted the wrong way and with negative intentions. Like all things that are good within this universe, it could be manipulated and twisted into something dark, and the lower-vibrational beings had found a way to do just this.

This time, however, was different from the last. As I felt the burning sensation throughout my being, I could not manage to open my eyes. Enveloped in complete darkness, I was nevertheless able to clearly sense the presence of the vile hooded beings all around me, as their chants grew in intensity and volume. To my great relief, I was abruptly saved this time by my mother's wakeup call from the concierge. Realizing that I was back in my physical body, I darted to the bathroom, frantically blasting on the creaky faucet my mother had complained about earlier

Hooded being in the chandi in Jogjakarta, captured on camera by the author.

that day and drenching my face in icy-cold water to ensure that I would not doze off again.

"Mom, we need to change rooms. This room doesn't have a good energy," I explained to my mother in a serious tone as I walked out of the bathroom, drying my face with a clean towel.

"Don't be silly. All the rooms are the same. It's just your imagination and your mind playing tricks on you," she replied dismissively. But as she peered at me, I noticed her expression change from one of ridicule to apprehension. "Actually, I had dreams of humungous snakes, too, creeping up my body. What does that mean? Should we change our room? But what if it's just our imagination? Should I call the front desk lady?" she asked worriedly, as she lifted both her hands to her mouth.

After my mother had managed to successfully request a room change for us and satisfy her sweet tooth by devouring a complimentary bowl of *es teler*, her favorite Indonesian dessert, she turned to me as she sat sprawled out on the hotel lobby's plush couch. "So, where would you like to go today?" she asked excitedly in her usual peppy manner.

"Hmmm," I pretended to stall and hesitate, although I had already planned out our excursion for that day. "How about the ancient ruins off the beaten path? The ones that tourists don't normally go to?" I asked enthusiastically, giving her an encouraging thumbs up. My mother didn't look too happy with my suggestion and looked down in attempt to mask her frown. She had probably already planned out her own day for us, centered on more eating and shopping, but nevertheless she gamely went along with my idea.

An hour later, we found ourselves in what my mother had described as "literally the middle of nowhere." Amid vast plots of fenced rice fields and scattered abandoned shacks, there was not a single person in sight. "Um, where are the ancient ruins?" my mother asked our driver, a timid man in his 50s, as she frantically whipped her head from left to right in desperate search for an official entrance. Unsuccessfully masking a chuckle, he pointed a slender finger toward a small *candi* (an ancient Javanese temple) surrounded by old, overgrown trees, "There, madam." She cringed in terror at the sight as I clasped my hands together in unbounded excitement and darted out of the car, gesturing for her to come along. She simply shook her head and pointed sympathetically at her beloved cell phone.

The *candi* itself was rather disappointing. Unusually small and completely dark inside, I stood inside it, looking around aimlessly and struggling to make out what was inside. As I was about to turn around and make my way back to the car, I suddenly felt a familiar pinching sensation emerge within my chest, the same uncomfortable feeling I get when roaming around in my energetic body in the lower-dimensional realms. Within a second, I found myself in the energetic. I appeared to still be in the same *candi*, except that, as I made my way around it, I realized that its interior was significantly larger than it had been in the physical.

I walked back toward the entrance, eager to see what the rest of earth would look like in this particular lower dimension, but before I could even manage to move an inch, I became completely paralyzed. My eyes were overcome by a stinging heaviness and were soon forcibly shut by a force that I neither understood nor recognized—that is, until I heard the same chanting resonate through my aching mind once again: "Ohm... ohm...ohm." The chants repeated over and over, as though they were echoes violently bouncing around in a confined space, as I struggled to focus on thoughts of love and light. Just as before, I felt my entire being getting burned slowly, inch by inch, straight through to my core. I fought against the pain and terror that threatened to overtake me, attempting to get myself into a meditative state and focus all my thoughts on calling upon the higher beings for their assistance.

I was soon lifted up and I felt the pain begin to dissipate. I slowly opened my eyes and found myself staring into a peculiar, white emptiness. *Have they finally managed to successfully cut me from my physical? Am I physically dead now? Am I disconnected from my physical?* My thoughts stumbled over each other in my mind.

"We laugh with sincerity upon hearing your thoughts. They can never do such," a soft voice echoed through my mind, soothing my panicking thoughts.

I craned my sore neck, searching for the source of the calming voice, until I stopped myself and closed my eyes: "How come I cannot see you, my higher beings?"

"We are within you," the voice communicated.

"My higher beings, can you please tell me who those hooded beings are? They've come to attack me multiple times now. What do they want?"

The voice remained silent for a moment before responding, "The consciousness has ceased to exist within them. It is in these ancient dwellings that were once used for positivity that they have focused their energetic dark force, and this has enabled them to pass through the earth's energetics. They remain around to spread even more darkness through earth. We have heard your thoughts. *Ohm* is indeed a form of energetic vibration that is pure and of a high vibrational nature, but when chanted in a certain rhythm, it is one that assists in the destruction of peace and purity—the soul's essence, or what humans call consciousness. They wish to evoke fear within you, to remind you of your human vulnerabilities. Energetic power is continuously increased by the forces that contribute to it, and this is why these dark beings gather in groups to perform their chanting. But you must not fear, for you are beyond them and we are within you."

At those last words I opened my eyes to a translucent, blue ray of light illuminating the vast empty space in front of me. It drifted toward me, enveloping my solar plexus with a cool, soothing sensation, and I was soon back in my physical inside of the same small dimly lit *candi*. I ran as fast as I could toward the car, the muddy soil shooting out behind me from underneath my sneakers. "What is the use of having a phone if you don't bother to answer it?" my mother angrily barked at me, as I swung open the car door and sank into the seat next to her. "We must hurry or we'll be late for the night performance of *Ramayana!*" I could see our driver's eyes widen in the rear view mirror. He quickly backed up onto the bumpy road and we were soon on our way.

Later on that night, half of the audience had left the auditorium within the first 15 minutes of the performance. Though it had indeed been a rather disheartening rendition of *Ramayana*, I insisted to my mother that we stay until the very end. Most people scorn Rama for obsessing exuberantly over Sita and, as a result, creating a war. Instead, I viewed *Ramayana* as a beautiful, symbolic story of the various forces of the universe— love, darkness, light, and harmony—at play. I saw the different energetic existences being depicted, how disharmony is created when those of darkness attempt to take the force of love away from those of light, and how, eventually, it would be harmony, love, and light that would triumph over the force of darkness. My mother, whom I explained my interpretation to later that night, could not come to see it this way: "I get it. I do. But I don't understand how you don't see the characters as characters," she quipped as she shuffled through her purse in search for our room key.

"Because, Mother, everything in this universe is an energy, a force. To see people only for their physical components is to constrict yourself to superficialities," I explained to her, to which she simply bobbed her head up and down.

"A-ha! Finally!" she screeched, holding up our room key as if it were a winning lottery ticket. A neon green light lit up over the card slot and we entered our room.

Formless Shadow Beings (Leeches)

After a seven-hour flight to New York City, I arrived at John F. Kennedy International Airport utterly exhausted, my disheveled hair looking as though it had been electrocuted. Katy, who has traveled with me many times, once commented in reference to my hair that "at least it sort of looks like Einstein's." I must be the only one to consider that a compliment.

My overall vibrational frequency that night was, I would say, average—neither very high nor too low—and my mind was at peace. After a hazy 45 minutes in the back of a yellow cab, I found myself checking into a hotel in SoHo, as my usual inclination for procrastinating meant that I had not officially signed the lease to my new apartment yet. It was during my night stay in New York's (supposedly) trendiest neighborhood that I would learn the incredibly crucial lesson of always following my intuition, no matter how inconvenient it may be.

As I shuffled tiredly into the hotel room, I was immediately struck by a strong feeling of a negative energetic presence. With an uncomfortable heaviness in my chest, I could tell that even my own vibrational frequency seemed to have slightly fluctuated upon stepping into that room. Despite all of these red flags, I chose to ignore my senses and instead convinced myself that requesting a room change, let alone a refund, and finding another hotel would simply be too much of a hassle. As I drifted into sleep, with small fractions of my consciousness transferring to my energetic body, I suddenly felt a painful, constricting sensation come from somewhere behind me. I was about to scream out in panic for help, thinking than an intruder had found a way into my room, but it was only after I noticed the, dark shadowy arm forcefully wrapped around me, that I realized that I was indeed in my energetic body.

Although I was shocked and frightened, I was blessed to have already faced such negative beings before. In fact, my numerous encounters with them had earned them an unpleasant nickname in my mind: energetic leeches. Energetic leeches, as I call them, tend to lurk discreetly around the human energetic. It is in the transitional state in which humans have yet to completely detach from their physical, that energetic leeches are able to successfully latch onto the human energetic. Their aims are malicious and straightforward: to lower their target's overall energetic frequency and to deplete them of their energetic fuel. As the majority of people remain unaware of their energetic body during sleep, these beings normally go about their business smoothly, encountering zero problems whatsoever. It is the small percentage of people who are able to automatically transfer their consciousness into full alertness during the state of sleep, who quickly realize the presence of these energetic leeches and attempt to escape them. In retaliation, the energetic leeches will either forcefully constrict their victims or use electromagnetic attacks—in many cases, both. After having encountered these dark beings countless times, I have come to realize that the best method to effectively counter them is to slowly elevate your own vibrational frequency in a discreet manner.

For this very reason, after detecting their presence that night, I remained still and in silence for about 30 seconds. It was during these 30 seconds that I noticed two more energetic leeches glide into clear view in front of me. I could thoroughly make out the outlines of their figures; they resembled people except that they appeared as black shadows, their frightening skulls completely bare of flesh. I immediately ordered my mind to focus on the feelings of high vibrations, those of boundless love and light. As I envisioned the powerful force of love flowing within me and resonating throughout my being, the energetic leech that had managed to latch onto me immediately removed its constricting grip from around my chest and fled back into the darkness from where it came. By rapidly elevating my overall vibrational frequency, I was able to break through the attacking mechanism of the energetic leeches.

The effectiveness of this method can be attributed to the universal law of energetic harmony, which prevents the extremely low vibrations of energetic leeches and the elevated vibrations of love and light—two contrasting vibrational frequencies of opposing rhythms—from occupying the same energetic space.

Nevertheless, energetic leeches have successfully learned how to vibrate at the same low frequencies as the majority of people, leading many of these people to remain entirely unaware that they are supplying these leeches with their energetic fuel every night. It's not uncommon to hear people say that they went to bed in a good mood but woke up in a foul one. Although mainstream scientists and psychologists have put forward umpteen reasons related to human physiology as to why this might be, it can be accounted for by the fact that we are all energetic beings living in a world of vibrational existences. Although most people live their lives focusing solely on the physical, continually absorbing and contributing to false, physical-based data formulated to divert humans from the truth, it is absolutely pertinent that you instead fully embrace and understand the reality of the universe and become an expert in the true essence of the world, that of the energetic. Make it a routine to always elevate your frequency to as high a state as possible before falling asleep and embarking on your energetic adventures. I will discuss specific ways to do this in the Helpful Tips section later on in this chapter.

Lower-Dimensional Entities

Though the majority of lower-dimensional entities used to be humans, most of them have lost their souls and failed to retain even the slightest amount of consciousness. As a consequence, they have come to exist aimlessly in the energetic, automated and confined to negativity and darkness. I have learned that it is only recently that these beings have drastically increased in number and acquired innate electromagnetic abilities.

While it is the right thing to do to show them compassion, the only surefire way to help them is through the elevation of our own level of consciousness. Because our foundation is energetic, when we collectively begin to vibrate in a very high vibrational frequency, it is likely that our elevated energy and vibrational frequency will automatically assist these lower-dimensional beings in heightening their own, through the natural law of the universe. It is through the elevation of their overall vibrational frequency that they will regain consciousness.

The operations

It was in the middle of falling asleep, the state in which we are most vulnerable, that I experienced yet another brutal attack. The perpetrators of the attack began as they usually did: sending me vibrational frequencies of the negative range to ensure that I would vibrate in a low rhythm while in the fourth energetic realm. Within an instant, I felt completely depleted of energy, a sensation that I could most closely relate to the dreadful time I had to go under anesthesia to get all four of my wisdom teeth pulled out. Despite never having experienced this exact sensation in any of my previous attacks, I had a pretty good idea of what was different this time and why. My attackers had entirely depleted me of the energetic fuel residing within my heart, a brutal method of attack that may have shocked me then but comes as no surprise now, considering the advanced technologies that have enabled them to engage in such dark acts with such incredible ease.

Although I once again lacked the strength to open my eyes, I used this knowledge as a clue as to where I was. Strapped into what felt like a dentist's chair, I could sense that I was in a laboratory of some sort. I had noticed these places in my travels before, often situated in areas such as shopping malls where the public would least expect them. Still blinded from the beings that surrounded me, I felt myself swivel around in the chair at their command. In that very moment, it became apparent to me that these dark beings had not only acquired the technology to see the energetic, but to also entrap it for their own personal use.

My thoughts were soon interrupted by an odd tingling sensation concentrated on the area of my third eye. It felt nothing like the electromagnetic shocks that I had experienced in previous attacks, and was accompanied by a jarring drilling noise in between my eyes. Though it did not cause me unbearable pain, I was well aware that allowing them to continue their "operation" would only enable them to destroy the connection between my mind and soul.

I quickly began to envision a high-vibrational energy of glimmering, yellow light begin to envelop my entire being, as I instructed my physical body to wake itself up. I threw myself forcefully off the edge of my bed and tried to get myself to the bathroom, but my energy had been so drastically depleted during this energetic operation that I fell asleep again and found myself powerfully vacuumed back into the lab by the same magnetizing force as before. Strapped into an operating chair yet again, I felt the operation target the center of my chest this time.

The wise higher beings had taught me well ahead of this particular experience that nothing can stop the will of the mind, and it was by focusing on this knowledge that I then managed to lift my left hand to my face and forcefully pry my eyes open, only to find a group of harsh lights hovering over me. The drilling noise faded into emptiness, and the operation stopped almost as if on cue, and the entities whose presence I had felt surrounding me scattered toward the corners of the large room so that I would not see them. Unable to move due to the intense exhaustion weighing down my body, through the corner of my eye I could clearly make out at least 20 identical operating chairs in the room. Lying on one of them was a woman who appeared to be unconscious.

"What you are doing is a crime against humanity, your own race and species!" I furiously shouted out into the expanse of the room, but received nothing other than my echo in return. I continued with evident frustration, "You are committing a crime against humanity. Do you not realize this? How shameful of you to do this. I will let people know the truth one day, one way or another!" It was with that promise that I felt the familiar throbbing sensation of an electromagnetic force being inflicted on my entire being. I directed myself to wake up in my physical body, and to my great surprise and relief found myself lying on the cold floor of my bedroom. Pushing off with my trembling hands, I managed to pull myself up and take a bath, for I knew that if I did not immerse myself in water to heal my energetic body, it would disconnect itself from my heart and I would be unable to travel.

Subconscious attacks

In the world of science, whether you ask a psychologist, a neurologist, or a biologist, you'll get all kinds of definitions for the word *subconscious*. For the purpose of this book, the subconscious simply refers to a blurry, somewhat undefined area that resides between the human energetic body (the unconscious) and the mind (consciousness). The subconscious marks yet another region of extreme vulnerability for humans. Unless an individual's overall level of consciousness is increased, the subconscious will remain an entirely incontrollable and hence susceptible region of the human energetic mind.

It is within the subconscious that one's unresolved negative thoughts, worries, and fears are left to dwell. When frequencies of the lower and dark range are targeted specifically at this area of the mind, the result is an annoying and often unbearable inner "chatter" of voices. Because most people do not possess adequate consciousness to remain aware and alert in their energetic body during sleep, as their level of consciousness is only sufficient to reside in the human mind, their subconscious will instead automatically take control and manifest negative environments and feared scenarios for their energetic body to go through—what many people refer to as nightmares. These self-induced nightmares, which deplete your energetic fuel, are what cause you to wake up after hours of sleep still feeling exhausted. With this in mind, it is extremely important that you first identify these subconscious "chatters" and then learn to shut them out before going to bed. Instead of catering to them and, consequently, allowing your subconscious to take charge, you must instead devote your thoughts to the forces of love and light, and focus on heightening the overall vibrational frequency of your energetic body.

Helpful Tips

Why are humans targeted?

If you are experiencing these attacks, then you are either one who has come to recognize the true knowledge or one who has unknowingly and accidentally stumbled upon the hidden truth.

Here are some reasons why these attacks occur:

* **To prevent you from discovering your true energetic abilities.** These dark beings are well aware that as more people realize their true energetic potentials, they will lose the power they have over humanity. Because we are a race that thrives when united and suffers when divided, the power of joint testimony is the greatest social force for assisting humanity in realizing its energetic abilities at this present time. By masking our abilities through attacks in the form of energetic torture that take place in the mind, these dark beings force us to brush off these negative experiences as figments of our imagination; what's more, due to the social pressures of our world, we are encouraged to keep these experiences to ourselves.

* **To inflict the fear that is associated with the unknown.** If you remain free from the emotion of fear and continuously seek the unknown, even while these dark beings target you with these attacks, you will undoubtedly fall into the whirlwind of truth and come to terms with your energetic abilities, just as I did as a child in my grandfather's pond. By inflicting the sensation of fear in people

through these attacks, the dark beings create avoidance within their minds of anything that is associated with the unknown, preventing them from further seeking the truth and elevating their amount of consciousness.

✳ **Torture.** Darkness never stops at level one; it continues on ruthlessly until its victims are rendered useless, hopeless, and dark themselves. No matter how big or small a crack may be, light will always find a way to pass through it. In this same way, it does not matter whether you have consciously traveled out of your physical body once or on numerous occasions, or whether these journeys have been epic or short-lived. These travels prove that you have let the light of truth come within you, and this makes you a threat to the dark forces.

Signs of an electromagnetic attack

✳ **Achiness.** Prior to being attacked, you will feel an aching sensation begin within your heart and then spread through your entire being—one that is reminiscent of the feeling you get after running a marathon or being sick with the flu. You must learn to identify and acknowledge it immediately.

✳ **Sounds.** Sounds serve as excellent indicators for when your energetic body is being tricked and forcibly pulled into the lower-dimensional existences. Upon experiencing these attacks, you will hear pinging, clicking, or clacking noises. If you have not gained complete control and alertness during your travels, you will not be able to retain any memory of being attacked and instead, will hear these kinds of sounds, along with a throbbing sensation of pain, upon waking up in the physical. By attacking you while your energetic body is still in the process of detaching from the physical body, these dark beings ensure that you will wake up almost immediately in a state of fear and physical pain. Because humans, in particular, are more vulnerable and accustomed to pain in the physical rather than in the energetic, attacks during the moment of detachment guarantee that the pain is felt by the victim.

✳ **A feeling of being pulled.** If you are a conscious energetic traveler, you will feel a magnetizing sensation, similar to the force of a vacuum, pull your energetic body into a lower-dimensional state when you are being attacked. This pulling sensation is caused by the unnatural lowering of one's vibrational frequency, which then results in the energetic body being forced into a lower dimensional realm.

✳ **Feeling drained.** As you drift off to sleep, you may suddenly find yourself feeling absolutely drained, as though the very life-force had been sucked out of you. This feeling is similar to being jet-lagged after a 24-hour flight. It is a crucial signal that must not be ignored. Such a sensation is the result of the depletion of one's energetic fuel that is caused by the lowering of one's overall vibrational frequency.

Signs that you were energetically attacked in your sleep

1. **Exhaustion upon waking.** All attacks in the energetic world are perpetrated with the primary target of drastically lowering an individual's overall energetic vibrational frequency. The first sign that your vibrational frequency has been energetically lowered is waking up in an incredibly exhausted state, despite going to bed in a normal or even energized state. The feeling of exhaustion is *always* caused by a decline in one's overall energetic frequency.

2. **Nightmares.** Most people do not possess sufficient con-sciousness in the energetic when they are asleep, and therefore do not remember being attacked once they are awake. Even if you have no memory whatsoever of being in your energetic body, the energetic is constantly being targeted, even when we remain unaware of it. This is re-flected through experiencing nightmares and remember-ing them when waking up in the physical.

3. **A bad mood.** When you experience a nightmare, but have actually unconsciously been attacked in the ener-getic, these attacks will alter your behavior and manifest themselves in the physical in the form of a negative mood upon waking.

4. **Negative thoughts and emotions.** After enduring an at-tack, the human energetic body will cease to exist in the fourth-dimensional human energetic realm, and instead will be automatically pulled down into an even lower di-mension due to the drastic lowering of its vibrational fre-quency. Any negative thought waves sent by beings that inhabit these dimensions will automatically masquerade as self-perpetrated thoughts within the individual's mind.

5. **Feeling disconnected.** Feelings of disconnectedness from the higher existences of the universe will be internally re-flected through feelings of emptiness and isolation.

What to do before, during, and after an attack

These tips and tricks have been written so that you are prepared to face any obstacles you may encounter during your travels.

✳ **Maintain a positive state of mind (before, during, and after).** Always maintain a very high vibrational fre-quency, a positive state of mind, and a positive emotional state. If you find yourself needing a mental or emotional boost, a fluffy companion such as a dog or an activity such as watching dolphins will always assist you in tuning into the force of love, which will then increase your vibrational frequency.

✳ **Use the power of manifestation (before and during).** Because these attacks take place in the fourth-dimensional realm, you will be able to use your power of manifestation to counteract them. By understanding that your thoughts govern both creation within this specific human energetic realm as well as the elevation of your own frequency, you will be able to manifest an energetic shield of protection—preferably of a golden-yellow color, as this is the color that resonates the highest energetic vibrations and thus is composed of the positive forces of love and light. Even when you are physically awake and functioning in your day-to-day life, your thoughts remain linked to the human energetic of the fourth-dimensional realm. Therefore, it is useful to use the force of golden-yellow light at all times, both when you are physically awake and on the verge of falling asleep.

✳ **Turn off your electronics (before).** As modern-day electronics have been strategically created to manipulate and lower our vibrational frequencies, it is imperative that you turn them all off, particularly cell phones and laptops, before going to sleep and refrain from charging them while you sleep. Electromagnetics are the real enemy of our true essence, which as you now know is composed of vibrations.

✳ **Embrace the power of words (before, during, and after).** Words are our most powerful tool for controlling the most vulnerable aspect of our mind, the subconscious. In order to triumph over the hurdles that the dark forces place in your way to sway you from your positive path to discovery, you must constantly remind yourself of the innate energetic abilities that are within you. To do this, you should focus only on those words that resonate positivity and vibrate in the forces of love and light.

* **Call for assistance (during).** Even when you are in the midst of being attacked, you will still be able to call to the higher beings of love and light for assistance. If you are unlucky enough to find yourself isolated inside a force field, the assistance you receive will be communicated through intuition. Instead of hearing a distinct voice instructing you on what to do, you will receive messages in the form of feelings that vibrate throughout your entire being.

* **Awaken the physical (before and during).** This is the safest and wisest route for both escaping and safeguarding yourself from these attacks. If you sense that you are about to be attacked or find yourself actually experiencing an attack, you must tell your physical mind to wake your physical body. In the case that you find yourself depleted of your energetic fuel (a condition that can trigger further attacks) upon opening your eyes, you should try to lightly pinch yourself so that your consciousness is entirely transferred into the physical. Pull yourself up onto your feet as soon as you can, do a few jumping jacks, and make sure that all of your electronic devices, specifically laptops and cell phones, are completely turned off and not charging.

Exercise: Energetic Treatment

Just as oxygen is necessary for the living physical body, the golden light is necessary for the energetic body. In the energetic world, particularly the higher-dimensional realms, the golden light serves as a constant aid in the upkeep of our vibrational frequencies. Because the primary foundation of the universe rests on the power of light, the true essence of each person's energetic body is expressed through extremely high frequencies, which are reflected through the color gold in the energetic world.

Each and every one of us possesses the innate ability to manifest the golden light and wield it as a powerful tool to assist us in raising our overall vibrational frequency. This golden light not only serves as a healing aid, but also as a shield against induced negative subconscious stimulations. Though you may at times find yourself constricted by your physical body, the golden light is an energetic tool that effectively counteracts the many vulnerabilities of the physical body. Envisioning the golden light, as described in the following exercise, will not only help improve your state of mind, but will also increase a greater sense of positivity and overall well-being.

Envision a shimmering stream of golden light flowing through the area in between your eyes, into your head, and then entirely surrounding it with its radiance.

Feel the rhythmic movement of the golden light as it illuminates and travels through your body.

Focus all of your thoughts and sensations on the healing power of the golden light and the heightening of your overall vibrational frequency with its every movement.

Feel the golden light eventually envelop your entire body, like a blanket of protection that will heal your every ailment and block all negative thoughts.

Note: No matter how sleepy you feel, it is imperative that you follow this exercise every night before falling asleep.

8

Harmony

'we never been much of a numbers person. Yes, I could quite easily and gladly go on about fractions of consciousness and tell you all about the varying realms and numerous planets I've been fortunate enough to travel to, but academically speaking, math has just never been a particularly enjoyable subject for me. However, I am and have always been extremely intrigued by numerology. Naturally, after discovering my own, I set out to find out why this was. I often wondered aloud to Katy why it was that every time I met a "five" (including her!) I always felt such a close connection to that person. Every description of my numerology always mentions this. Deprived of an answer, I turned to various sources to find clues. *Time is a powerful determinant of individuals and their personalities*, nearly all of them claimed. But time, as you very well know by now, is merely a manmade fragment of the human imagination. It gives us an artificial sense of harmony, when the reality is that the universe itself is the ultimate determinant of that.

It was not long after embarking on my travels through the energetic realms, that I began to encounter numerous constraints on a regular basis. Even on occasions when my overall energetic frequency was vibrating in an elevated level, I would find my travels unjustifiably limited by restrictions, which, as I would soon discover, had been implemented as yet further attempts to prevent people from traveling multidimensionally. On one such night, I remember knowing that I was going to be automatically pulled into the human energetic realm before I could elevate myself to a higher-dimensional existence. Sure enough I found myself in the strange confines of my own bedroom, though in its energetic iteration.

The Monk

Upon separating from my physical body, I immediately noticed a figure swaying back and forth in a faintly lit corner of my room, its gray outline jumping from wall to wall with its every movement. Convinced that the figure was that of an earthbound spirit, I began to approach it cautiously as I dedicated my thoughts to elevating my vibrational frequency so that I could assist the earthbound through the conjoining of our rhythms. (I've learned through experience that this is the quickest way to do this, though I do not recommend attempting this if you are a novice.)

The figure turned out to be that of a short, slender woman who was wearing a long, light brown robe with a hood that covered her head entirely. Once I was standing in front of her, I smiled warmly, hoping to elicit the same genial hospitality I have received during so many of my travels. Closing my eyes, I lifted both my index and middle fingers and gently placed them together on the side of her neck, continuing my attempt to assist her in raising her vibrational frequency. But when I felt the unexpected warmth of her body under my touch, instead of the usual coolness of earthbound spirits, a sense of confusion overcame me. *There, directly in front of me, stands an earthbound*, I thought, *yet somehow she seems to be in her physical.*

"I'm not an earthbound, Khartika," she interrupted my thoughts with a dimpled smile. "Our vibrations intersected—or, you could say that our currents collided with one another because we had the same vibrational frequency. Although it may seem that this intersecting of our vibrational frequencies was merely an accident, it was very much intended. A fated occurrence, a fated meeting, a fated interaction sent by someone you know who is currently in the higher realms."

After feeling out the situation for a few moments and failing to detect any negativity or deception within her, I politely asked her who she was.

"I will raise our vibrational frequencies as we speak so that you can travel to meet this being," she explained, not exactly answering my question in the manner that I had hoped.

"Please do excuse my inquisitiveness, but I am extremely curious: Do you exist in the energetic or the physical?" I asked her as I sat down on the edge of my bed.

She walked slowly across my room, with her long, flowing robe drifting behind her. She reached a wall and stared intently at a taped-up poster of my favorite quote by Albert Einstein: "The most beautiful thing we can experience is the mysterious," it read in chunky, bold letters across a backdrop of the frothy, blue swirl of the Milky Way. "I exist in the physical," she replied casually, as though there was nothing unusual about it. She quickly turned from the poster to me as though she had suddenly uncovered a distant memory, her robe twirling to keep up with her, and continued, "How are your parents? Is your father all grown up now?" Her question caught me entirely off guard, like a heavy rainstorm during a stuffy, sun-swept, California summer day. I wondered how she could possibly know my family, for though they may be religious, they are certainly far from spiritual.

"Yes, he's still short-tempered and bad-mannered," I answered hesitantly, still rather suspicious of her relation to my family.

She chuckled and exclaimed joyfully, "Well, it's his nature, isn't it? His frequency never matches yours, so it is no doubt that harmony does not exist between the two of you."

She soon began walking toward the opposite side of my room, this time moving her conjoined hands peculiarly across the wall, as though she were praying. Somehow, I knew that she was trying to rid my room of any negative energies that lurked within it. "You know, there exists a blueprint of human energetic frequencies. Here on earth, they calculate it through numbers, but these numbers are actually basic vibrational codes that one's energetic being vibrates to. You mustn't be one of those humans who ignores this and then wonders why your relationships are tragic. These numbers serve as a bridge between the physical world and the energetic world, and studying them further can assist in the maintenance of harmony in the human energetic world. They are, in essence, the physical representation of natural occurrences in the energetic world. There are two numbers—two basic energetic vibrations, really—that are most important, and you must learn to recognize these two vibrations. They are the sevens and the fives," the being explained as she proceeded carefully around my room, continuing her energetic cleansing against the smooth wall.

"Yes, I'm aware of numerology, but I had no idea that these numbers are energetic blueprints that assist in maintaining harmony amongst humans. The sevens and the fives... Can you please tell me more?" I asked her in an almost pleading tone, expressing my boundless interest in the subject.

"You are a seven. Sevens are most harmonized with other sevens, or with fives. In fact, in the energetic world the vibrations of two sevens make up a five: seven plus seven equals 14, and one plus four equals five. Five is the energy of inertia, of movement, of happenings. Two sevens together will create great change and new knowledge. Seven and five make three: seven plus five equals 12, and one plus two equals three. In the energetic sense, three is the number that humans view as a symbol of good luck or blessings; your culture, in particular, has come to call it *fa choi* (good fortune). Why do you think there is a saying on earth that everything happens in threes? In terms of our understanding, it is not really luck, fortune, or blessings; it is just the right harmonic vibrations that join together and resonate energies that create further energetic positivity or sequencing. You must remember these two energetic blueprints because the right friendships in your life will vibrate in sevens and fives. They are the ones whose basic components, when joined with yours, will create energetic positivity. Though they may be scared by the knowledge you will share with them, it will speak to their souls and resonate in due time."

I looked down at my hands that were folded neatly across my lap. Something still didn't make sense to me, and there was no way that I could leave myself hanging, so close yet so far from the knowledge and truth that I sought. "But why specifically would sevens and fives accept the truth about our energetic worlds; why them in particular?" I pondered aloud.

"Every being's existence in the various dimensions, as you know, is fated to occur. As fate is very much determined by natural events, these numbers that are used, which humans call numerology, bear a significance that [has to do with] to energetic bonds. And to answer your question, these two imprints of the energetic vibrations of sevens and fives—their essences are not those of earth," the being responded in her soft voice.

Intrigued, I instantly blurted out, "Wait! Does this mean that every being has been vibrating in the same basic blueprint for many lifetimes? So...it is permanent and does not change through lifetimes? I thought that life-path numerologies change over lives."

At this, the being made her way toward my bed and sat down next to me. The hood of her robe slipped back slightly and for the first time since the beginning of our conversation, I was able to see her eyes. Though they were brown like my own, they exuded a vibrancy that I cannot describe, moreover entirely comprehend. It was as though they had been illuminated by the knowledge she was sharing with me; as if, alongside me, she was learning and discovering it for the very first time. "It is very much applicable across lifetimes, as this is a natural occurrence of your being within the universe and many universes, and this being, as you know, persists through many lifetimes. But for now, I can only say that those that are sevens have known ever since they were young that they are not like the others. They do not think like others and they are well aware of this. They've acquired much wisdom from the many incarnations they have had; they are tired, worn out, and want to leave this earth and go home. The fives, on the other hand, are very new to earth. It is either their first or their second incarnation here, so it's no surprise that they are still true to and in tune with their energetic existence. Though they are what humans would call the 'new kids on the block,' they are very sensitive people, extremely attuned to other peoples' feelings and emotions, and that is why some can't understand why they are this way and unfortunately try to get rid of this side of themselves through external means. Of course, the reason behind this is that they are so in tune with the energetic world that they feel things strongly, but they do not know how to cope with [that].

It is a sad reality that many of them living on earth try to self-medicate through worldly pleasures, mostly due to feelings of depression that they can't seem to comprehend or find reasons behind. It is incredibly unfortunate that most of them do not come to understand that this depression is due to feelings of loneliness and the awareness that they are not of earth. So, unless a five learns of spirituality and the truth behind their existence, many of them have a high chance of resorting to these external dangers of the physical world. You will also come to notice that fives are more attracted to music than most other vibrations. This is because music, especially music that is electric in nature, transfers rhythmic frequencies to their energetic bodies and, as a result, makes them more in tune with the energetic world. For them, this produces an inexplicable feeling of satisfaction because they are unconsciously aware of their true energetic nature! Do you see? These basic blueprints exist in all humans, yet many of them cannot come to understand why some people repel others! You must remember that it is the sevens and the

fives that will give you energetic harmony in this physical life, so friendships with sevens and fives will be beneficial, as they are instinctively
aware of the energetic worlds."

The being paused suddenly and looked at me intently before pro-
claiming, "I've cleansed your room; someone will meet you here very
soon."

I sensed that she was preparing to leave, but I had one more question
for her: "Do you mean to say that anyone who is a seven or a five will be
a great friend to me in this physical life?" I asked quickly, begging the
universe to not end our conversation there.

"That may be true, but you must remember that these numbers only
relay the very basic vibrational frequency that one's being vibrates in.
Everything is taken into account in determining a person's overall vibrational frequency—not only his or her birthdate, but also his or her
birth time, name, state of emotions, and the energy he or she reflects
into the world; all of these factors determine your overall vibrational
rhythm as well as what harsh obstacles you will face in this physical life.
Try thinking of a piece of metal covered in rust and corrosion. This piece
of metal is still metal, isn't it? It is only the oxidation and the external
environment it has been exposed to that have caused the metal to appear this way. The rust and corrosion can either be scraped off or left to
accumulate. You must realize that it would not be wise to assume that
all [sevens and fives] come from positive or high dimensions, for, as you
have learned, there have been many alterations throughout the universe.
Now," the being quickly pulled herself onto her feet, her brown robe following obediently behind her. "In Togetherness...someone is waiting to
speak to you." She looked at me, her brown, almond-shaped eyes twinkling with that vibrancy, as though she were expecting me to respond,
but I could only manage to think of my subconscious worries. *What if I
get pulled back?* I wondered frantically. *What if I wake up before it's time?*
These unbearable thoughts flooded my subconscious, isolating me from
the being standing in my room. *This idea of numerology—people in our world*

will probably think that it's too "out there" for them. But it is as real as the genetic code that makes up the physical body. It maps out the probabilities of the universe; it is a language of the energetic world. It would be like us telling the people of the 1800s about inheritable genes that are activated through interactions with the envi—

At this, I was abruptly pulled back into my physical body, terribly disappointed in myself for my lack of control over the recurring, subconscious chatter that chased itself through my mind. I knew that I now faced two choices: I could either return to my sleep and attempt to find the being, or I could wake myself up and accept that it was simply not meant to be. After considering these two options briefly, I settled on the latter. It was the kind being herself who had wisely told me that this world functions on the basis of energetic harmony, and that everything occurs only in the right moment. Serendipity. Harmony itself is the true universal meaning of what we call fate. Who knows? Perhaps fate will allow us to cross paths again. Since this fateful travel, I have wondered long and hard about this being and who she might be. And through all my sleepless nights, one thing became undeniably clear: She was a citizen of the higher realms.

The Land of Koo

After a year of consistent traveling, I had yet to come across an energetic being whose outlook could enlighten my mind further. I had indeed learned of the fundamental aspects of energetic travel. Every night, as others slept and refueled for the day's work ahead of them, I found myself lending my trust to the universe, hoping with all my heart that it would lead me to experiences that would further my knowledge of the energetic worlds.

On one of these nights, after successfully detaching from my physical body, I found myself soaring through my apartment building at an unfathomable speed, my upstairs neighbor's spoiled Chihuahua, Spotty, yipping and yapping at me from his special orthopedic cushion as he watched me disappear in the distance over him and into the vast universe. I felt my chest constrict in mild panic as I watched Spotty become a faint dot, followed by my tall, skinny apartment building, and then the planet we call home. (By the way, despite what anyone may tell you, traveler or not, you don't just instantly become fearless in the energetic world. Never having been entirely comfortable with heights, I still feel the subdued panic whenever I find myself flying through space, though much less so than if I had been riding a rollercoaster, for I know that no harm can befall me in my energetic body.)

I closed my eyes, shifting my entire focus to the universe, and suddenly felt the direction of my energetic body shift into a horizontal course. When I opened my eyes, I was in a dimly lit bedroom. In spite of the poor illumination, I saw that it was meticulously neat and organized, a far cry from my disorderly room on earth. Overflowing with my usual curiosity, I began to quietly find my way around when I heard the deep, disembodied voice of a man: "Hello? Who are you?"

I introduced myself in the most straightforward manner I could manage: "My name is Khartika, earth, energetic traveler, year 2011."

As I turned around to face the voice I had heard, the room instantly lit up as though on the man's command. I was incredibly surprised to see that the man standing in front of me was Asian; I had never encountered an Asian in any of my previous multidimensional travels. I could now see the room in which we stood more clearly. There was a white, wooden bed positioned in the center, with a nightstand to its side and a closet in the corner. I assumed it was the man's room. "You exist in the physical, don't you? Do you exist in the Milky Way galaxy, too, like earth? Is your planet the same size as planet earth?" I inquired curiously.

The man gazed intently at the dimmed lamp resting on the nightstand next to the bed, and as he began to speak, I saw it flicker brightly and illuminate the walls around us. "Yes, we do. But this is not the Milky Way galaxy, and we are much smaller than earth. I've always known that a visitor from earth was to visit us soon. You are most welcome. I have a lot to show you." He quickly grabbed a cream-colored trench coat from his closest, threw it over his shoulder, and led me outside the room as the lights dimmed behind us.

A few steps and we were standing side-by-side in front of the door that we had just exited. I could not believe that I was on a different planet. Our surroundings were strikingly similar to the suburbs of Japan, which I had seen numerous times on the *shinkansen* (bullet train). In the distance of green expanse that stretched across the horizon ahead of us, I could see numerous towering temples adorned with the intricate, sloping rooftops typical of traditional Japanese architecture. Thrown off by these observations, I gazed curiously at the man; he was around 5 feet tall, with a robust build and short, dark hair. There was nothing unique or particularly alien about him; he had the demeanor of a Japanese professor.

"I really do not understand how this can be a different planet. It looks exactly like the country of Japan on planet earth," I told him, a bit of frustration apparent in my tone.

He looked at me sympathetically, as though he longed deeply for me to understand, before affirming with a light sigh, "Khar— Ti— we are not on earth. This is not earth. We are of a physical planet and of a different galaxy than earth. Yes, I am familiar with earth. I had a life on earth. But what differentiates our planet from earth is that rather than focusing on the physical, here we focus on the energetic. We are aware of the truth. Have you noticed that everything around us is deserted, empty? That you have not seen any people at all?"

"Well, no," I admitted. "But they could be sleeping or just inside their homes," I suggested.

He seemed amused by my answer, as he chuckled deeply and scratched his head. "That is a very human answer, Khar! There are certain periods of our cycle, or what you humans would call a week, that we devote to aligning our minds and resonating in high vibrations to travel around the universe and communicate with the higher beings. Everyone participates."

Though I was slightly offended by the man's initial comment, I knew that his intentions were good. I asked him why I couldn't see the other residents of the planet, considering that we were all in the energetic.

He remained silent for a bit as I followed him through a park of scattered bamboo trees amid puddles of fluffy, golden forest grass. "We are communally aiming for the sixth dimension," he said. "And that is where they are currently. Energetically, you are in the fifth dimension right now."

I was both confused and surprised by the implications of what he said. "So, you've managed to rise as a planet energetically, and, as a result, all of your energetics exist in the fifth dimensional existence, as opposed to the fourth, which humans go to when they sleep? When you sleep, do you exist in the fifth?" I asked, while marveling at the vibrancy of the bamboo trees.

He nodded, bringing his hands together in front of him in agreement. "Yes, it is possible. We once thought that it wasn't possible, but our achievements proved that such wasn't the case. It is through numbers, through unity, that we were able to achieve it, Khar. At first, only a mere few of us managed to do this, to come to exist in the fifth dimension. But as we gradually increased in number and our entire population joined in, we all began to energetically exist in the fifth dimension."

We exited the beautiful park and stepped out onto a clean, narrow sidewalk, strolling down its path together until I felt the man gently place his hand on my shoulder. We had finally reached our destination—a long, single-level, white rectangular building composed entirely of concrete. It reminded me of a gigantic Lego piece. I passed through the glass door in the front of the building without any difficulty, finding myself in a spacious, brightly illuminated lobby. It was completely devoid of furniture and people, as though it had been intentionally abandoned long ago. "Welcome to planet Koo, Khar!" The man's orotund voice rang out into the building's emptiness, and with these words, I saw the floor of the building transform, Escher-like, into at least 30 unfathomable levels of descending escalators that intersected with one another. With an inviting grin, he guided me toward one of them and we were soon moving deeper and deeper into the entangled maze.

"I really don't understand. It doesn't look like any apocalypse has affected your planet outside. Why do you live underground? Is it for safety?" I asked as we descended down the escalator.

"It is for safety. Our outer planet is just for show. We travel undisturbed in the underground. Traveling—humans still call it sleeping, don't they?" He peered down at the seemingly never-ending escalators below us and I followed his gaze, but his answer had not satisfied my overwhelming curiosity.

"I don't understand how you have buildings that look so similar to those on planet earth, and how this can be the physical. I can't make sense of your planet at all."

He lifted his head up and let out a soft chuckle. "Khar! You do not actually believe that the higher beings that traveled to your planet, particularly the ones that influenced the building of what humans know as temples, only came to earth, do you? I expected more of you, Khar. But never mind about it all! Come! Let's see the exciting things. You don't have to move so rigidly. Glide instead. Glide! You're in the energetic—glide!" I watched my new acquaintance move effortlessly onto a neighboring escalator, joining into the rhythm of the moving steps. I attempted to follow his graceful movements and was quickly astounded by my success. *At least I'm not as clumsy in the energetic as I am in the physical*, I smiled to myself.

"Stop!" the man suddenly yelled, jolting me away from marveling at my new achievement, as he leaped onto the platform of a passing floor. I followed quickly behind him, jumping up as high as I could so that I wouldn't get left behind on the escalator. We had reached a space that reminded me of a laboratory, and we were surrounded by long, wooden tables on which aquarium tanks were scattered. These were filled with metallic objects immersed in translucent water. "Exciting, isn't it?" the man exclaimed as he approached one of the aquariums, rested his elbows on the table, and gazed intently at what it held. "Do you know what you're looking at?" he asked, sensing my presence next to him. I shook my head and he pointed to a metallic cluster, about the size of a coin in diameter, that looked like a magnet. "These are the elements of the universe; the people of your world have only discovered a small *fraction* of what you see in this entire room! We have found that these elements, when reacting with another force, can either enhance or disrupt thought patterns in relation to energetic traveling. We place them in water to determine which effect each element has. That outer world you see, it's just a show. We are fortunate enough to have found solace within the interior aspects of our world. And even in our components and our very own being, it is the interior that we have focused on and continue to thrive in. Though there are many outer characteristics of our planet that are similar to your planet, we choose to look within instead."

As his words found purchase in my mind, I felt a knot form in my heart. "I do not like my planet. Is there any possible way that you could find me a physical here? Is there any way that you could just keep me here, please? I'd like to be a resident of this planet," I quavered, my speech breaking, on the verge of tears.

He turned away from the aquarium and looked up at me with regret in his eyes, explaining apologetically, "It is not permitted, though after your journey ends on earth you could always opt to come and live here permanently. Please listen to me. We fight and continue to triumph over darkness; water blocks the negative, lower-vibrational frequencies that target our planet. Humans must know to focus internally. If the fate of your planet is the same as ours, they must learn that the exterior makes them vulnerable to low frequencies. The one message that I'd like for you to always remember is that harmony must be maintained amongst those of like energetic frequencies. It is through harmony that *it* will be spread. Focus on those who focus on the internal."

While listening to him speak, I once again began to worry that I would be pulled back to my physical body at such an important moment. I also knew that such worry would very soon lower my overall vibrational frequency to such an extent that it would be a self-fulfilling prophecy. "Visit me! I'm getting pulled back now! Visit me! Assist us!" I called out to the man, as I watched the aquariums dwindle away and woke up in my physical body.

Center's Moon

My travel to Center's Moon had not come easily. Early that night, I had found myself dreadfully stuck in a nightmare induced by my own subconscious fears. One after another, subconscious scenarios rooted in various personal fears launched themselves out of my mind and into my dreams in a seemingly endless cycle. I was forced to follow along as my grandmother slipped in a puddle of water and was hospitalized. Everything I saw seemed so real, even more so since the dream included someone I was so emotionally attached to. Unable to detect the unreality of these scenarios, my mind had been terribly deceived, and my consciousness had stayed within my physical, leaving my energetic utterly uncontrollable.

It was not until I had noticed an obvious oddity in the dream—my dog was a different breed—that I finally realized I was in the human energetic realm. My consciousness was automatically transferred to my energetic body, enabling me to rightfully take charge of it once again. While envisioning the powerful golden light spread itself over my entire being, I focused on aligning my emotions to the forces of love and light, increasing my energetic fuel, thus enabling myself to manifest flight and entirely break free of the shackled human energetic realm. *I align my thoughts and vibrations to the frequency of the higher dimensions. I wish to travel to a planet of the higher dimensions*, I declared telepathically, trusting the force of attraction within the universe to lead me to the right planet.

"Welcome!" A message zipped through my mind in the form of an energetic frequency, like an impatient taxi driver in New York City, as I found myself being guided through the vacant interior of a vacuum-like tunnel and into the depths of an ocean on a planet I had never visited before. I felt the invisible force gradually slow its pace, as though I had reached my fated destination, then suddenly jerking me into a wide, spacious corridor with unusually high ceilings. Everything around me was of a light mahogany color and rustic design, reminding me of a peaceful cabin in the middle of the woods. Ahead of me, the corridor appeared to continue infinitely in a smooth, orderly course. I thought about following it, but was soon distracted by a breathtakingly beautiful, melodious soprano voice traveling gracefully toward me from the farthest side of the corridor, singing, "Center's Moon, Center's Moon, I'd like to go to Center's Moon." With every syllable that resonated, I felt my energetic frequency being elevated to a rhythm that matched that of the melody. I obediently followed the silvery, captivating voice, making my way across the dimly lit corridor, which to my surprise curved smoothly into a distant corner. I found myself walking right into a crowd of distracted beings gathered around the area where the voice seemed to be coming from. Everyone's eyes were fixed in a collective gaze toward the center of the formed circle, their hands clasped together directly in front of their foreheads.

Unable to catch a glimpse of the singer from where I was standing, I braced to push myself through the growing crowd, when I suddenly felt a hand on my shoulder and received the telepathic message to follow it. Led by my innate inquisitiveness, I trailed behind the being, a petite woman with long, sandy blonde hair, who appeared to be in her 50s, and I immediately sensed a nurturing energy about her. "Hello. My name is Ana. We are of the ninth-dimensional existence," I heard her assert in a composed manner, as we turned sharply and entered a separate wing of the corridor I had not noticed before.

"My name is Khartika. Pleased to meet you, Ana. Were you the one to welcome me when I was being led within your planet?" I asked her curiously while shuffling behind her.

"You were speaking to the collective force of this universe, the force of Togetherness." Though I did not understand exactly what she was referring to at that time, I pretended to nod in agreement, as there were many other questions I desperately wanted to ask her.

The corridor soon gave way to numerous small, empty rooms opening from both sides of the hallway, each one of them missing a door. As we passed by them, I saw Ana's eyes skim quickly through their unoccupied interiors. She seemed to be searching for something. She gradually slowed down her pace so that I was walking directly next to her. "I heard a being earlier," I said. "She was singing beautifully about Center's Moon. May I ask you a question about the existence of moons? What is the purpose of a moon?"

The crevices of her mouth smoothed out and formed into a grin, her eyes eliciting a sense of peace that I could feel deep within my heart. "Moons often exist in conjunction with planets that have had beings moved out and created the moons." I nodded at her response, attempting to comprehend it and relate it to the existence of our own moon, until the being suddenly stopped in her tracks and turned to face me. Gazing at me, she lifted her right hand and rested the tip of her forefinger gently between my eyes. I immediately felt a cool surge of energy travel smoothly from the back of my head down my spine, invigorating my being, as though it were a freshwater stream in the middle of a rocky canyon. "You desire to know the truth about those individuals whom physical beings of earth call 'children with mental disabilities,'" the being communicated telepathically to me, addressing a question that had long lingered within my mind. "This term is incorrect, as these individuals are actually energetically conscious beings. Their natural beings exist in a way that does not allow them to perfectly align their energetic with their physical in the same way that most humans do. Although this is often viewed as an imperfection on earth by humans that deem themselves to be perfect, this is simply a result of a misalignment between the physical body and the energetic body."

Still feeling the cooling sensation, I smiled at her explanation. It confirmed a truth that I had long suspected, and I was instantly inspired to ask for further knowledge on the topic. "What can I do to assist these individuals on earth, Ana? What advice can I give them?" She did not respond for a while, as we walked side-by-side through the long hallway in absolute peace and silence. It was as though she were focusing on something deep within her to provide me with the correct answer, and it was in this moment that I felt the strong sense of sympathy, understanding, and—above all—love within her for these individuals on earth. Standing next to her, I was overcome by an unexpected wave of love and light, and for what seemed to be the very first time in this current life of mine, I felt as though I were one with not only her, but all those individuals who have so often been put down by others on earth.

In what seemed to be a few earth minutes later, the hallway began to widen unexpectedly, its walls stretching outward and welcoming us with their open arms. Eager to see where it would lead us, I glanced ahead gleefully, squinting my eyes and quickening my pace with every stride. We found ourselves standing in the center of a round room entirely enveloped in a dome-shaped wall of glass; outside we were surrounded by aqua blue, crystalline water in every direction. As I marveled at the extraordinarily beautiful sight, I heard Ana's soft-spoken voice echo within me, "The message that you are to give these individuals is that they should never succumb to negative thoughts, those thoughts that deem their physical bodies to be weaker or more vulnerable than those of so-called 'ordinary' human beings. We must encourage them to take part in activities such as what humans call meditation, which is particularly focused on raising vibrational frequencies and allows them to find peace of mind. During this state of positivity, love, and light, they no longer experience this misalignment. It enables them to understand that they are one with something bigger, and that they are beings of positivity, love, and light. But perhaps the most important exercise for these individuals to take part in is that of the golden light. They must envision this golden light flowing in from in between their physical eyes, and then spreading through the body like a golden blanket of protection. They must be reminded of and recognized for their energetic abilities, and they must know that these energetic abilities certainly do not have anything to do with their resemblance to other human beings. Their energetic abilities are far above anything that can be manifested in the physical."

I was burdened by an undeniable feeling of excitement. "So, these individuals are energetic travelers! This means that if I assist in raising the awareness of the truth behind their abilities, they should be able to travel much more easily than most!"

Ana nodded, placing her hand gently over mine. "This is very much so! Your statement brings me joy. However, it is important to understand that traveling does not always occur consciously for all of these beings. Although the misalignment between the energetic body and the physical body does make these travels more likely, you may wonder what has caused this. It is a result of many of these individuals being energetically forced to exist on earth in a way in which they do not necessarily align entirely with their physical bodies."

I nodded in response to her explanation, but my confusion must have been evident, as she soon continued, "You can very much relate this to something being forcefully taken from its environment and placed into an entirely different environment. All of these differences in the new environment, much like temperatures and weather on planet earth, make the pursuit of life and existence in the new environment incredibly difficult for the being." Ana paused briefly, looking intently at me as though expecting more questions. "These beings were taken from their original dimensions of existence and forcefully placed within the physical," she answered, picking up on my flustered mental frequency.

A thought instantly entered my mind, and I blurted out, "And when these beings are forced to exist here on physical planet earth, their high vibrational frequencies attempt to assimilate themselves to vibrate in the same low vibrational frequencies as those of the majority of beings. So, is it because their nature goes against this, that they experience this misalignment? This makes sense for many mental disabilities, from predominantly inattentive ADHD to autism."

"That is very much so. Because they were not able to properly alter their vibrational frequencies, their vibrational frequencies have been skewed and drastically misaligned to the point that they feel connected with the vibrational frequency of their original dimension and the realm they originate from sometimes, while [they feel] entirely disconnected at other times. As humanity is a race that demands knowledge (which, as you know, has often been provided to them in a skewed and falsified manner), it is important to note that there is very clear evidence of this. All of their 'symptoms' prove that these individuals are very much still in tune with their energetic selves. Many humans often categorize those children who talk to themselves when no one else is present as having mental disabilities, but this is very much clear evidence that most of these individuals' consciousness exists in their energetic selves. In realms where beings vibrate in elevated energetic frequencies, communications occur through energetic, or, what most humans call telepathic, means. Therefore, when these individuals experience misalignment and are placed into physical bodies, they are forced to voice their communication out loud, as this is the human accustomed and conditioned way for doing so. However, they still attempt to communicate in the energetic way, as this is what feels natural to their energetic being, only their communication is now exposed in the human conditioned way. As these individuals have been conditioned to act this way due to the barriers of planet earth, they must be taught and allowed to connect to their energetic selves—in essence, be conditioned in the opposite way."

After hearing Ana's explanation, I immediately asked her about the many children with mental disabilities who are unable to relate to the emotions of others or read other individual's facial expressions. "Why is this so?" I wondered aloud, furrowing my brow in confusion. I could tell that Ana had already sensed my question long before I had asked it.

"Though they have been conditioned to act within the barriers present on physical planet earth, they have also not entirely adapted to the physical and are very much still in tune with their energetic selves. Thus, they cannot identify with these human aspects, as their nature solely enables them to sense emotions and feelings through energetic frequencies."

I stood still for a while, mulling over Ana's words, while she watched the undulating water slip and slide against the glass over us. As she hummed to the same captivating melody I had heard upon my arrival on the planet, other beings began to walk by us in small groups, smiling warmly at us as they passed. I felt an odd sense of familiarity within my being, almost as if I had a connection with them, until I found myself thinking of my grandmother, who currently suffers from Alzheimer's disease. I asked Ana longingly if she could give me more knowledge about this.

"This is very much a disconnectedness of one's energetic being from their physical existence," she said. "As you know, the human physical body is constructed in such a way that it gradually ceases to function appropriately over the human time scale. This disconnectedness of the energetic body from the physical body is a natural process that occurs for the safety of the energetic body and being. In such cases as your grandmother's, it is the separation from the energetic body, when the physical body begins to deteriorate in its functionality, that permits energetic traveling to occur. Although it may seem to others that these beings do not have control, they actually do indeed have control in the energetic sense. It is only in the physical sense that they seem to not have control, as their energetic being has detached from their physical body. They do indeed retain consciousness of their energetic travels and they could very much travel wherever they desire, wherever in the universe they intend to, as long as they are able to perceive the true abilities of their being."

With her words fluttering rapidly through my mind in desperate hope of being retained, I was overcome by a powerful sense of relief. In that very moment I finally felt at peace with thoughts of my grandmother, understanding for perhaps the first time that though her health may be drastically declining in terms of the physical, she is now gaining even more control and consciousness in her energetic, something that I have strived for since the beginning of my travels through the energetic realms. I asked Ana how I could help my grandmother and others like her.

She gazed up into the blue expanse of water that drifted over us, and said, "Primarily, you must recognize how crucial it is to not think of these individuals as being mentally unstable. These individuals can greatly benefit from the attempts of those who surround them to discuss their travels with them. Of course, these travels need not be referred to as 'energetic travels.' Nevertheless, asking [them] about the places they have visited would certainly be of immense assistance to them."

I focused on securing Ana's advice deep within my consciousness, so that I would remember it upon returning to the physical. "Ana," I exclaimed. "But how should I communicate with them, as they do not understand language in the same way, particularly those with drastic autism?"

"This should be done through using visual methods. For example, simply being in the presence of water can assist them in achieving the elevation of their vibrational frequencies. This is no different than what would be recommended to any being currently experiencing their journey in the third dimensional world. It is through visual [means], such as demonstrating that a drop of water connects to and unites with an entire body of water, that they themselves can realize that such unity applies to them, too. Symbols, as we sense that you know, have long been used and portrayed by numerous ancient civilizations, such as that of the Minoans. They serve as energetic portrayals of the physical, and thus would enable beings who have experienced misalignment in the third dimension to connect to their abilities and this idea of unity. Positivity often functions in the presence of simplicity; thus, these particular symbols need not be complicated or of an elaborate nature. Water is certainly a symbol of positivity, as is a ray of light that enters through a crack and then spreads and disperses itself through an entire room."

Taking everything in, I slowly looked up and noticed a glimmer of light quickly making its way through the iridescent water that surrounded us. It embraced us with a glowing intensity, unmatched by that of the sun on planet earth, and I knew instantly that it had been sent from somewhere deep within the universe. I smiled and asked Ana my final question, fated by the universe to conclude our meeting. "These children, especially low-functioning autistic children—they are forced to adapt to a society of 'normal' humans, or else they can't live a 'normal' life, as humans put it. I wish, in this present life of mine on earth, to make a difference and to help them. What can I teach them, Ana, so that they can become experts in the energetic worlds?"

I felt her answer resonate instantly through both my mind and my heart. "You must spread knowledge. This is the primary task which you must endeavor [to complete]. The spreading of knowledge will illuminate the recognition and the truth behind these beings. You must remember that these individuals should be encouraged with symbols, metaphors, and meditation, so that they can increase their vibrational frequencies."

I opened my eyes to find myself back in my physical body, still feeling Ana's peaceful, illuminating energy within me. After managing to pull myself out of bed, I shuffled over the cold tile floor to the living room, where I could hear that my little sister had brought her Singaporean boyfriend over. Sprawled over pillows on the ground, they chewed their lunch together over the coffee table as they stared intently at the television screen, watching one of their many favorite TV shows. I smiled at them and continued on to the kitchen to make myself a hot cup of tea, the beautiful melody that I had heard earlier still playing faintly in the back of my mind. As I hummed to its peaceful rhythm and skidded around the kitchen island to fill the water boiler with one smooth sweep of my right arm, I heard my sister's boyfriend say to her in a tone of warning, "*Aiyo!* Dear, please don't eat fried stuff. Later your face get worse, *lah*. More pimple come out." There was a brief, pregnant silence, before I heard her bark back at him, "How dare you? When your face had that episode, did I say anything? And it was way worse than mine! Who the hell says that to their girlfriend?" Through the corner of my eye, I saw him take an especially large bite of his sandwich in attempt to divert eye contact with her. "Sorry, sorry," he mumbled, in between chewing, for the sake of maintaining harmony.

Helpful Tips

Energetic fuel

Energetic fuel is produced and elevated through the constant alignment of one's thoughts and emotions with the forces of love and light. Energetic fuel is produced within the heart and then distributed to the mind, where it is used by the energetic body.

Though in differing amounts, every being in this universe possesses energetic fuel, as it is an essential factor of our existence. Energetic fuel is what enables the heightening of one's overall vibrational frequency and hence serves as the "powerhouse" for our travels through the energetic realms. As the mind is interrelated with the heart and thus governs the energetic body, it is absolutely pertinent that you maintain thoughts of positivity and feelings of pure love and light so that there is enough consciousness present within your energetic body for achieving conscious travels.

Joint interaction

In the higher-dimensional worlds, the conjoining of rhythmic forces is thought of as a natural occurrence. This perpetual interaction between forces, which elicits an elevated joint rhythm, further enhances the harmonious nature of all the universes and is rooted in the concept of Togetherness, in which vibrations merge together to maintain the high state of energetic positivity in the higher realms. Much like two narrow streams that merge together into a powerful waterfall, when vibrations interact to form a united rhythm of oneness, their energies are elevated to an enhanced vibrational frequency.

When it comes to true attraction here on planet earth, the attraction we feel is in fact energetic in nature, as we are naturally attracted to those individuals who increase our overall vibrational frequencies. When our vibrational frequencies are elevated, the energetic rhythmic functionality of the heart is also increased; this is the true meaning behind popular endearments such as "You are my heart," proving that most of us are aware (albeit unconsciously) of our energetic essence.

Unfortunately, such knowledge has been drastically altered on physical earth to elicit negative low frequencies instead of positive high ones. As a consequence, rather than accepting the true energetic nature of our being, most humans have turned to indulging in the act of joint rhythm for the sole purpose of feeling chemical pleasure in their physical bodies. The people we call "sex addicts" are in actuality individuals who vibrate in low overall vibrational frequencies and partake in sexual acts in attempt to increase their frequencies; because they do not possess the true understanding of joint rhythm, their actions lead to only temporary increases in their vibrational frequencies, which more closely resemble "energetic stealing." With this in mind, it comes as no surprise that many sex addicts experienced childhoods that were deprived of love and affection.

Those individuals who seek to achieve joint rhythm in the absence of love are essentially attempting to find a shortcut to elevating their low vibrational frequencies. Sadly, these actions have been further triggered by the vulnerability of the human brain, which has been strategically conditioned to desire physical pleasure. When individuals indulge in joint rhythm for the simple purpose of attaining pleasure (entirely devoid of love), their actions will actually have the opposite effect, lowering their overall vibrational frequencies and pushing them further from their spiritual evolution.

It is absolutely crucial that you begin to see every thought, action, and situation from an energetic point of view. Whenever you find thoughts that are driven by animalistic desires arising in your mind, your must remember to envelop yourself with feelings of love and light so that you can elevate your overall vibrational frequency.

The mastery of flight

Individuals who have achieved flight in the energetic realms are often referred to as *teleports*. Although teleports remain a rarity in the fifth- and sixth-dimensional realms, each and every one of us can become one. This is because the ability to manifest flight is dependent on one's ability to reach an elevated level of vibrational energetic frequency. Nevertheless, if your reasons for wanting to teleport are selfish and negative, you will never be able to learn how, no matter how disciplined you are.

The ability to fly in the energetic realms is incredibly different from the act of simply directing yourself to a particular destination and then finding yourself there. Instead, the mastery of flight relies heavily on the amount of energetic fuel that is stored within your heart. While the mind, much like the steering wheel of a car, dictates direction, it is the heart—the fuel—that stores the necessary energy. In order to achieve flight you must possess extra fuel, as this will enable you to go beyond the norm of average travelers of the energetic realms.

Contrary to simply traveling in the energetic, which relies heavily on the mind, the ability to achieve flight in the energetic depends on both the alignment of the heart with the mind, as well as the presence of immense feelings of love within the heart. Though you may be accustomed to seeing everything in terms of the physical, the feeling of love fuels your energetic body in the same way that food powers the physical and protein creates muscle.

Finding Togetherness: Part 1

"This place exists on the power of Togetherness. It would not be possible for me to tell you... It's all Togetherness. And the only way I can tell you and you might be able to comprehend it, is that we are all together and keep all of them moving," the great being said to me.

"What is this place?" I asked, fairly trembling in pure joy, as though my soul were confirming my whereabouts. Drenched in the planet's raindrops melodiously trickling and splashing against me in a rhythm, I waited patiently for a response from the great being. With every word I communicated and every gentle nudging of the rain, I felt a warmth of positivity traverse excitedly throughout my being, increasing my frequency at a pace I had never imagined possible.

"Love, light, and positivity—those were the birthstones of our universe. Earth's creation was not one that occurred through the natural occurrences of the universe, unlike that of the other planets. Its manifestation into the universe was entirely unnatural. Earth was created by a group of beings. Though they were initially good beings, they gradually degenerated into those of negativity and darkness.

They used and abused their powers, gaining even more knowledge of the forces of the universe as they traveled to the different realms and planetary existences. Everyone welcomed them, as we did here, generously showering them with great knowledge of the various forces of the universe. One cannot even come close to imagining how many different forces exist within this universe that can be used by good beings to do wonderful things. Instead, these beings misused their newly attained powers and manipulated various forces in this universe to create earth."

I sensed a distinct sadness within me, an emotion that could never be mistaken for another, and that always managed to leave a mark on my heart. It dug into my being, searching for something that had been buried, something that was meant to be uncovered. And as its search continued into the crevices of my being, sadness was soon overtaken and outshined by the zeal of liberation—the liberation of finally knowing the truth. I had long chased aimlessly after this knowledge, misinformed that the history behind earth's creation was something tangible, something that could be captured. Having received this information, I thought of the earth I had known while growing up, a disproportionate ball of seemingly captivating green and blue that twirled through space on its slanted axis. Immersed in this vision, I wondered curiously about these beings that had created earth, and if they were all of the same race and species.

"Though they originated from a variety of different dimensions and clashed in the universe, interestingly, they were all entirely unique and did not conform to one particular race. This is what made the prospect of creating a highly identical race particularly entertaining and an attractive option for them," the great being's voice resonated within me, as though the space that separated us had no meaning. "After they were overcome by darkness, their negative energetic frequencies were attracted to one another, creating a clash."

Looking off to one side, I caught sight of a vibrant, golden-toned horizon, masked in the watery haze of the rain. The rain moved peculiarly, traveling sideways instead of down. It poured continuously against me, as though welcoming me, but I knew that its purpose was really to assist me in maintaining my vibrational frequency.

"They came to our planet along with many others in the higher realms, and, unaware of their true nature and intentions, we willingly gave them the knowledge of such forces. We had assumed that their ability to travel to our planet was the result of their elevated vibrational frequencies, not their deceptive technological advancements. As there was no clear information on these dark technologies, we did not know of the possibility of artificially altering energetic frequencies; thus, we deemed it to be correct and in accordance with the foundation of the universe to accept these beings as those of love and light.

"You must have heard of the concept of evolution. Although the topic itself is often a subject of controversy, the process of evolution did occur on earth, but not quite in the same way as it is regarded by humans of today. Ironically, these beings of the negative range evolved over time to be that of the human nature. However, not all of them evolved completely; a small number of these beings experienced a backlash. Outnumbered, these beings were ousted—forced by the humans to leave planet earth.

"Prior to earth's creation, there was already an existing moon, only one of a much smaller size. Using the same forces they created earth with, these beings soon turned to expanding the moon so that they could escape the opposition that they experienced on earth. They deemed that by residing on the moon, they could think of ways to further manipulate the universal forces to alter earth, undisturbed.

"Interestingly, some of them, though very few, decided not to flee to the moon, and instead opted to exist somewhere within the earth. This was their last resort, one which would allow them to remain both distant from the human race, while at the same time close enough to resume communication and still be able to watch over them."

I thought that I had known everything, only realizing now that my knowledge was but a few drops in the ocean. I gazed into the sky above me, earnestly seeking the familiarity of the moon, but failed to spot even a single one. Its absence struck me, and my deep confusion was likely palpable. I found myself jammed inside of a narrow maze of perception, its path paved with the asphalt of misconceptions I had picked up on earth. *Who were these beings presently residing on the moon, then?*

"These beings do very much resemble humans, but most certainly their physical appearances have been altered. You can think of this as an evolution to better adapt to the atmosphere and environmental conditions of the moon. The beings that exist in the physical dimension on the moon only inhabit the inner region of the moon, while the darker beings that inhabit the outer [part of the] moon do not exist in the physical dimension.

"The humans who have visited the moon have not been in communication with these physical beings, as they have not entered the inner moon and have had no prior knowledge of the existence of beings in the inner moon. These darker beings, who migrated to the moon from earth, have been utilizing earth's sources of water to support the beings that exist in the physical dimension and live inside the moon. You may be curious about the [evidence of digging] that has been observed on the moon by humans. None of these observations is attributed to the existence of any particular mineral on the moon; they are simply pathways that darker beings who do not exist in the physical dimension use to transport themselves and communicate with those beings that live in the interior of the moon."

"On earth there is a theory that these dark beings came to earth to acquire gold. Is that true?" I asked.

"It is certainly a misconception, as the spread of darkness and the attainment of evil is very much the primary goal of these dark beings, so this should not be considered true by humans."

In the midst of trying to rearrange these puzzle pieces of knowledge, one still seemed out of place, its indecipherable edges stubbornly refusing to fit into the others within the confines of my mind. *I suppose this means that every moon is a residence for dark beings.*

"Not in particular. Moons exist for a variety of different reasons, not simply [to serve as homes for] certain beings.... It is unwise to assume that all beings that reside within the moon are dark. Every moon is distinct in its own way. Though there most certainly are moons that have been created to serve as additional residences for the beings of a planet, moons can also serve an energetic purpose, functioning as an energetic portal.

"Once the true existence of earth and the story behind it was revealed to all in the universe, there were certain beings of an entirely different nature, those of even greater darkness, that adopted terrible ideas from the actions of the dark beings who created earth. They formed even stronger agendas to spread darkness, visiting earth [secretively] and affecting human beings in ways that would permit them to quickly spread this darkness. These beings originated from a nature of evil and darkness and stemmed from the lower dimensions, their existences occurring in the universe as more and more darkness spread through it. These are the beings that are currently doing terrible things to the human race. With evil comes even greater evil. And though these two groups of beings appear to be different from one another, they are both conjoined together by the collective force of darkness."

Still somewhat attached to my human mind, I envisioned the many different forces of darkness as depicted by humans; but much like a Polaroid picture exposed to light, it soon faded into a gray fuzziness, as a stream of vibrant light, seemingly out of nowhere, unexpectedly flooded my sight. My gaze devotedly traced its luminous path backward, flurrying after it in swirls until finally falling upon a community of lilies. Their presence invigorated me, with their light illuminating my entire being. Every energetic particle of mine, as though moving in a rhythm to an unheard melody, captivated not only my heart, but my being in its entirety. My mind, a train that was heading to a false destination, was immediately halted.

"In very many cases they do not possess any precise form. Some of them exist energetically. Others incarnated to take on the human physical appearance. They most certainly do have the power to take over humans' physical bodies.

"They are aware that earth possesses a large population of humans that have been overcome by strong ignorance; though they possess this knowledge and the universal abilities, the majority are far too ignorant to recognize them and participate in the Togetherness. Hence, the main goal of these negative and evil beings is to indoctrinate the human race and population, a task which at this rate appears to be quite simple and relatively easy for them to achieve.

"These humans that are indoctrinated can be easily employed by these negative beings to assist them in venturing through the universe and attaining additional powers to further manipulate the universe, in ways that you and I have not even imagined possible.

"They intend for human beings to join them on their quest for manipulating and acquiring all the power, forces, and knowledge of the universe, as the population of humans on planet earth is by far exceedingly greater than the population of these beings.

"As the human civilizations of the past outnumbered and triumphed over them, they failed.... But as the group of even darker and evil beings began taking over the human race, this prompted those that humans refer to as the 'creators of earth' to once again manipulate and acquire power and control over the human race through falsely providing them with a sense of positivity."

Surrounded by towering stalks of glistening golden luminosity, whose brilliance and beauty not even the most elaborate and poetic depiction could bring justice to, I knew immediately that this place surpassed even the labyrinthine imagination that could be comprehended through one's physical mind, for such an attempt to convey this realm in the limited human comprehension would result in failure. The word *love*, in its purest and most uncorrupted form and meaning, is the only word from our human dimension that could describe this realm and all its beings.

My sight fell attentively upon the shimmering, golden stalks, swaying to and fro in unison to a peaceful rhythm. *There must be a wind that is causing this,* I immediately thought.

"They are not controlled by a breeze but by us; and when we shine, we exert this force," the great being's message resonated within me in the same beautiful rhythm as the golden stalks.

"Forgive me, but in my small mind, I cannot seem to comprehend what you are. For example, I am a species known as human. What are you composed of here, my great being? Also, may I ask, what is your age?"

"We smile at your curiosity. You may think of us as magnets of energy. We function on and identify various energies, and we communicate through energies because the force of Togetherness is, in fact, composed of energies. Here we do not chart or attempt to chart our existence, as unlike many humans, we know that our existence here is infinite. We strive toward the ultimate unity, perseverance, and continuity of the existences of Togetherness."

"Are you the creator of this universe? Or rather, do you believe this universe has a creator? Sorry—I do not even know what I'm saying. I suppose I am just at a loss for words. Never before in my travels have I ever ventured to such a place of a much higher existence."

"We function on the power of Togetherness," the being replied. "We are all united and we are all one. Here, we do not perceive an individual creator of the force of love and light, and thus we do not value any creator above us. We do not see ourselves as below; we are all one and united under the concept of Togetherness."

All one and united. Like the sun that illuminates the darkness on earth, these words illuminated both my heart and my mind.

"The cyclical nature of the higher dimensions of the universes thrives primarily on high energetic frequencies, and all who are connected to the power of Togetherness contribute to this cyclical existence through vibrating in the higher rhythmic frequencies."

Togetherness? I attempted to unravel the meaning behind this word.

"It is wise to think of the force of Togetherness as the positive unity that is fueled by thoughts of love, light, and positivity. It is the foundation and the building ground of the universe, and the various different universes and galaxies, and the higher dimensions and planets that they are composed of, and it is what permits them to continue persevering. It is the primary force of the beings of love and light, and it is what permits the universes and the higher dimensions to exist as they do. The power of Togetherness is bound to benefit anybody of a higher frequency and state of being that exists."

I watched as a petal of a lily detached itself and floated to the ground, a solid ground that glittered in golden-blue colors. I looked at it with great admiration, hoping that it would teach me its secret. As though welcoming my presence, the lilies created an invisible bridge with my being, made known through the harmonic melody that I both felt and heard in that moment. *Why, oh why, a beauty so true, an essence so pure, yet appreciated on earth only for its looks.*

"We are overjoyed by your acknowledgment of the lilies. Lilies do very much also pose energetic capabilities, and it is in the higher dimensions that many planets and many existences possess the presence of lilies. Since their formation in the universe and the many universes, lilies have contributed to the raising of higher vibrational frequencies."

Thoughts that had previously been clouded by the concrete, the physical, the tangible, began to precipitate in a rainfall of understanding, falling as if to the tune of the universe. Behind every note and melody that resonated through space, stood beings of high vibrational frequencies that partook in creating a single joint harmony—an energetic symphony to assist in uplifting the universe's overall rhythm. *Such must have been the relationship between the Minoans and lilies.*

With the clouds now faded into emptiness, my heart smiled at its new realization. The Togetherness in itself is a symphony, a symphony composed of varying rhythms and vibrations, all working together the betterment of a positive universe.

"These individuals, who you refer to as the society of the Minoans, were a civilization that greatly stressed the maintenance of energies. They very much lived in harmony with nature and, as humans would say, they were agricultural. It is this connectedness with nature that very much contributed to the maintenance and enhancement of their energies. As you very well know, this particular civilization differed drastically from [that of] the humans currently occupying planet earth. The Minoans dedicated many centers [that] specifically catered to the maintenance of energies; it is within these centers that energies would be maintained and enhanced before finally being transferred to other regions. As we sense that you are already aware of the power of water, you might find it interesting that their ability to maintain high vibrational frequencies was through the attainment of Togetherness with water. The Minoans also lived in harmony with other beings; they were aware of the energetic potentials and abilities of varying species, particularly bulls. Interestingly, the horns of a bull, during the time of the Minoans, were able to emit various vibrational frequencies, hence contributing to the overall elevation of other beings that lived in harmony with them.

"And those beings who did not necessarily come from extremely low dimensions, but came from dimensions lower than the physical human realm, such as snakes, were not frowned upon and instead assisted by the Minoans in the elevation of their vibrational frequencies. They were aided energetically in the heightening of their frequencies so that they, too, could live together in harmony and assist in the elevation of the overall vibrational frequency of their civilization, and, in essence, that of earth.

"However, some individuals within their society were eventually exposed to superficial distractions by outside influences that were indirectly affected by the forces of darkness, and this led to the destruction of a harmonic and energetically conscious society."

A feeling of admiration washed over me, as I reflected on my vibrational frequency, which continuously heightened itself by simply being in the great being's presence. "Is this the reason why there has been evidence found of human sacrifices taking place?"

"Some did occur, but it was only long after the darker forces infiltrated their society. Before such, the concept of Togetherness and energetic harmony were successfully maintained.

"Once their society was infiltrated by various sources of darkness, they had to resort to finding particular areas in the natural landscape of their region that were, during that specific period, more discrete, and hence less known by the darker forces.

"These areas often manifested as caves, and it is within these caves that the Minoans devoted most of their time to traveling energetically. And such was the only way that those individuals who maintained positivity, love, and light were able to evade the darkness that had infiltrated their harmonious societies."

The message was clear: The Minoans had looked upon the physical world that surrounded them through an energetic lens—a lens that enhanced all that the universe offered them, into a vision so pure and clear. Darkness, ever so innovative in its ways of corrupting the vulnerable human mind, lured them into fallacious perception, blurring their sight with an incomprehensible dizziness that shunned them from the truth they had once known. But it is not through the eyes that the truth is discerned; instead, it is the perseverance of the mind that triumphs over the darkness it has been dimmed by, and everything in the blur blends together into an inevitable whole.

I could see a trail of flickering energy navigate itself from within the depths of the group of lilies. It whirled gracefully through the glimmering expanse of the planet in a tone of creativity, until halting itself and forming a circle that was divided into four equal sections. With this energetic illustration, I felt a deep sense of irrefutable familiarity evoke itself within me: *I've seen that symbol before.* "My great being, I've seen many Minoan symbols. What exactly is the significance of a circle divided into four equal sections?"

"The Minoans were a society that very much focused on and emphasized particular symbols that reminded the individuals and beings living in their society of their energetic capabilities. It is very much like the propaganda of today's world on earth; such images can be used to either heighten the vibrational frequencies of beings and individuals that are living in the physical world, or to do the opposite and lower them.

"The encircled cross is very much a symbol of the fourth-dimensional realms. These beings living in the past ancient civilizations, such as the civilization of the Minoans, were energetic travelers, and this symbol of the fourth-dimensional realm, depicted by a cross split into four equal sections, further emphasized and highlighted the fact that in their society, they were not solely bound to the third-dimensional realm, and that there was the possibility of traversing into this fourth-dimensional realm, [a realm in which] they weren't necessarily bound infinitely by the physical human body."

"What about the symbol of the swastika that can also be found in numerous Minoan sites? May I ask, what does it signify?" I asked.

"It very much signifies an intersection of vibrational frequencies originating from the varying dimensions of the energetic world. In essence, the vibrational frequencies branch out from the center, where they interact in unity, to the outside, where they exist in their own dimensions. It is an energetic symbol that depicts the constant maintenance of harmonic unification."

It all made sense—an energetic representation of harmony inverted into something that was to be used instead to spread darkness on earth. Darkness, in its ever-evolving creative ways, had yet again managed to reverse something so pure and so powerful in our world, sinking it to the depths of the ocean like a heavy stone weighing down a beautiful origami boat, leaving it uncovered and forgotten on the sands of the seabed.

"We sense the desperation in your being. Though darkness and its evil doings have found its way to reverse all that is pure and good, there is a symbol that is known in your society, [one] that has been brought from the energetic into the physical by the society of what yours has come to call the Minoans; an extremely powerful symbol that even if inverted, would still convey the same positive meaning of the true essence of our universe. A symbol that conveys the same meaning when looked at from any angle, inverted or tilted, unconditionally."

Though I did not see it, I knew immediately what the great being was relaying to me, as though a projection of sensation was intentionally made to connect itself to my own. *Phaistos disc! What is the meaning of it, then?*

Replica of the Phaistos disc.

"It is very much a portrayal of the universal unification of the higher and positive dimensions, stressing that even the beings living and bound to the physical body in the third dimension do, in fact, possess the ability to become one and united in the force of Togetherness. As you can see, this image and its portrayal can be interpreted in numerous ways. It can be viewed as beginning in the center and then spreading outward, or as beginning in the outskirts and progressing inward.

"This depiction emphasizes the true essence of the force of Togetherness—that though this universe is composed of various beings of unique vibrational frequencies and natures, stemming from different planets and different realms across numerous galaxies, we are all—at the same time—united.

"It is impossible to look at any particular part of this image and fail to notice that it is a united whole, and this is exactly what [they were attempting to stress].

"A variety of symbols are repeated throughout, serving as symbolizations of the various dimensions that have existed in positivity, love, light, and unification, since the formation of the universe and the many universes.

"Though darkness currently exists within this universe, these symbols serve as proof that the dimensions of love, light, and positivity are destined to be forever united in the form of Togetherness, triumphing over the force of darkness.

"It is proof that a group of individuals encompassed by the force of love and light, vibrating in a high rhythm, can join in as a unitary force to further assist in the uplifting of the vibrational frequencies of the universe.

"In essence," the being concluded, "this is an energetic invitation to all positive beings in the universe. We are all on various different paths in the universe that beautifully fit into the concept of Togetherness. It is for the betterment of Togetherness that we are all on these particular paths, each and every being there is. Uniqueness instills the power of Togetherness."

While on my far-flung quest for knowledge, I had stumbled upon countless ladders. I accepted their invitations willingly, scaling them as if they were steep mountainsides filled with the promise of adventure, only to find that none of them descended past the typical logicalities of planet earth.

But their wisdom engulfed me like a powerful rip tide, swallowing my curiosity and everything I had previously thought and known in its ever-abiding wake. I no longer wanted to swim parallel to the shore, as I had been taught for as long as memory can reach, but instead against the tidal pull of the unrelenting current, so that I could remain captured within in. It is here that the universes collide, every drift hailing from disparate yet united sectors, joining together into one, like the determined bubbles of sea foam. To understand and to finally sink to the depths of the ocean of truth, created a warmth felt deeply within my heart.

10

Finding Togetherness: Part 2

Humans must know that Togetherness really does exist, and it can only be achieved through the power of the mind. Humans must be increasingly able to recognize and accept that such a power does exist, and it is through their own abilities and within their own minds that they can enhance and join in—in the power of Togetherness. Humans think that they possess such a vulnerable anatomy that could easily be hurt, and this further allows them to be weakened. It is crucial that they learn that their true essence does not even have to do with the physical aspects of what they deem to be their anatomy, and that their true power lies within themselves and their mind.

—The Togetherness

I watched a second petal from the lily detach itself willingly, as though it understood the unspoken means of the universe, a timeless existence for all beings.

If only the universe could make you see beyond the depths of the sky, beyond the clouds that meet the eye, to the horizon far and high, then with a single heartbeat, you would know that space and time are but cheap illusions that humanity buys.

I asked the great being: "I do not understand why most humans are the way they are, chasing after temporary gain and living their lives over and over again, entirely satisfied. I am tired of the ways of the physical world, whereby superficialities and material aspects exceed the mind; for one's physical body is one that is soon to be discarded, like the mountains of clothes in the rich man's closet. Why can't most humans realize this?"

"You must not think of yourself as being higher than everybody else. Although you might be wiser or have more knowledge or be of a more positive nature, you cannot think of yourself more highly in these aspects. Because it is part of everybody's path to discover his or her true energetic essence; most just do not notice it because they do not have this knowledge."

"Yes, my great being; you are correct. Thank you for pointing that out. It wasn't arrogance that I meant to convey. Please forgive me in my way of thinking."

"This is important because in the story of earth that was shared with you, the primary reason why the humans were able to so easily and quickly conquer the dark, negative evil beings in earth's past history, was that the negative beings had become what you call...arrogant."

"My great being, these dark individuals have implemented electromagnetic weapons to attack the human energetic body, though the militaries have masqueraded such technologies and claim that they are only used on the human physical body. I cannot understand the true mechanisms of such weapons. Please, if you can explain them to me I'd be most grateful."

"It is very much the goal of the Togetherness to stress the energetic over the physical, as such is the true essence of this universe," the being responded. "You must not think of things in terms of the physicality, as their aim is the energetics. These electromagnetic attacks have been focused on the modern technologies of the world; the darker beings have manipulated them in an orderly fashion as an aim to lower the overall vibrational frequencies of their targets. Such can be implemented in a variety of ways, the most common of which occurs through the electrical currents that naturally shoot out from these modern-day technologies. These technologies, such as laptops and other electronics, have the capacity to break down other electrical and rhythmic currents. Thus, when people vibrating in high vibrational frequencies (and hence emitting high rhythmic energy) use these technologies or are in the vicinity of them, the electrical currents of these technologies are attracted to these particular vibrational frequencies. It is their natural inclination to break them down, and what occurs is the lowering of the individual's vibrational frequency.

"As beings of a positive nature do constantly possess a certain amount of consciousness in their energetic, even if they have not, as humans say, *fallen asleep* yet, when they are attacked, these electromagnetic weapons send frequencies of the negative range that break up their high vibrational frequencies and thus remove the amount of consciousness that is present in their energetic body. When this consciousness is removed from the energetic body and transmitted back into the physical body, the connection between the physical and the energetic is disrupted. Though these attacks are directed at the energetic body, because the individual no longer possesses enough consciousness in the energetic body, he or she will feel a tingly sensation in the physical (as the energetic body still remains within the physical body). It is in this way that these electromagnetic attacks are translated into sensations and feelings in the human physical body.

"As these vibrational frequencies are broken up and consciousness is transferred out of the energetic body and into the physical body, the individual's vibrational frequency is lowered due to these electromagnetic frequencies of the negative range. This lowering of vibrational frequencies, in effect, damages the heart and causes fear within the human physical mind. The more often these attacks occur and the more the heart is damaged and targeted, the less potential the heart has of forming energetic fuel, which further prevents this individual from being able to elevate his or her vibrational frequency to a high level in the future, transfer consciousness into the energetic body, or retain consciousness while traveling energetically.

"Thus, the more these attacks are implemented, the more the heart is damaged and the more potential it has for losing its ability to produce energetic fuel. These electromagnetic weapons are being developed as the darker forces multiply and become increasingly aware of the existence of positive beings. They are well aware that some individuals have discovered how to break free from the constraints of the physical body and elevate themselves to a higher vibrational frequency, and these energetic weapons would enable them to better prevent this from occurring."

A thought occurred to me: "Are the individuals who create such advanced technologies, such as cell phones and laptops, beings of the dark force, then?"

"That is not necessarily so, as very many humans remain ignorantly unaware of these technologies. However, it is very much safe to say that due to the manipulative capabilities of these dark beings who form part of the dark and evil collective, it is very much within their abilities to connect with humans that have connections to these technologies, or even to take over them to promote this [agenda]."

I suddenly thought of a chaotic street in Jakarta, its pavement swarming with people and abused with litter and pollution. On either side I could make out countless utility poles, complex webs of tangled electrical wires scrunched desperately around every one of them. And much like these utility poles that support the jumbled stockpiles of cables, innocent people are unknowingly creating invisible networks in the human energetic realm, unaware that by turning on their modern electronic devices and leaving them on throughout the night, they are contributing to the system of electromagnetic networks that lowers our vibrational frequencies.

"Yes," the being said in response. "It is important to note that these attacks take advantage of the linear process—the formation of the energetic fuel within the heart, which leads to the elevation of vibrational frequencies and then the transferring of consciousness from the physical into the energetic self. They have taken advantage of this vital process in human beings, which is certainly pertinent to the existence of the energetic body, reversing it to go about attacking humans in brutal ways."

"My great being, can you please explain the relationship between energetic frequency and energetic fuel? I understand that energetic frequencies must be heightened in order to achieve conscious travels, but the energetic fuel that is within the heart—is it there to assist humans in recharging while they sleep? Is it what causes them to feel recharged upon waking up?"

"The energetic fuel and energetic vibrational frequencies of beings are very much interrelated. Energetic fuel is not related to sleep, as all beings have energetic fuel and require energetic fuel to exist, but not all beings necessarily require sleep. It is very much only an aspect of humans. Energetic fuel is what is produced by these feelings of love and light, and what in essence has the ability to increase vibrational frequencies and elevate them. Vibrational frequencies on their own cannot be increased, and it is the energetic fuel that serves as the powerhouse."

"May I ask what truly happens in the energetic sense when I find myself being transported to an operation table by them?"

"Ironically, though these attacks appear to be drastically different from one another, many of them are instigated against humans in highly similar and typical ways. They first target the human consciousness in the energetic body, as the dark beings understand that this is the essence of all beings that exist within the universe. Through using these electromagnetic weapons that evoke vibrational frequencies and energies of the negative range, any consciousness residing within the energetic body at the time is broken up and separated and thus degraded, so that all consciousness in the energetic is transferred back into the physical.

"Therefore, because no or little consciousness is left to reside in the energetic body, as a result of these weapons, humans who experience these attacks feel as though they are experiencing them in the physical, even though these attacks are very much conducted in the energetic and targeted at the energetic body."

"My great being, why do these weapons target even those humans who do not retain consciousness in their energetic?"

"These weapons target the vibrational frequencies of humans, and although a number of humans do not necessarily vibrate in especially high frequencies, they can still be put to use as targets for the darker forces. This breaking down and lowering of vibrational frequencies occurs primarily while the human is roaming around in his/her energetic body, and can occur both unconsciously and consciously. You can think of these weapons as working backward. When the human is targeted in the energetic body, his/her vibrational frequency is lowered to such a low point that it negatively affects his/her energetic fuel and thus lowers the amount of consciousness that is present in the energetic body. So even if this individual did not originally have consciousness in the energetic body, this consciousness will be diminished to such a point that it becomes locked into the physical body. Thus, from that point on, the attacks that they experience will be manifested as fear, pain, and negativity in the vulnerabilities of the human physical body, and this will furthermore succeed in preventing energetic fuel from being formed within the heart. What results is an ongoing destructive cycle in which these humans are unable to create energetic fuel due to these attacks, and hence cannot heighten their vibrational frequencies and find themselves unable to willingly transfer more consciousness into their energetic body."

Another question sparked my curiosity, but the great being answered me before I could ask: "It is a variety of beings, existing in different realms, both in the physical and the energetic, that are instigating these attacks."

"But I do not understand why researchers, even college students and some members of the military, are perpetrating these attacks on me and others? Surely, not all of them are overtaken by darkness," I said, in more of a question than a statement.

"It is very much because they are all simply directed to do so by others and provided with rewards that would be of benefit to them, such as the material concept of money. Many of these individuals remain unaware that they are instigating these attacks directly. It is very important to note that many contributors to these attacks, especially those who do not possess knowledge of the truth behind these so-called experiments, participate in these attacks in a variety of ways. Many of these attacks are conducted as group efforts, in which these individuals do not necessarily deal directly with the energetic body. Most are left blinded to the truth, with the false knowledge that they are simply conducting research or experimenting on frequencies and energies. However, a select number of them have become aware of their wrongdoings."

"Do they target all vibrational frequencies?" I interjected.

"They target any high vibrational frequencies they sense and come in contact with. Thus, their attacks target a wide variety of individuals. It is a different experience for you because you are able to retain memory of these attacks."

"Am I correct in saying that their desire is to use humans as fuel? And that, unknowingly to these humans, when their vibrational frequencies are low and negative, they are contributing to this force of darkness, even creating an energetic connection between themselves and the force of darkness?"

"That is very much so; as you have discovered, humans possess many vulnerabilities and inabilities simply due to the existence of the physical body. The physical body is the largest and most discouraging limit; as you may have discovered through your travels, human beings are being used, as you have termed, as fuel, or simply as soldiers of the army of darkness.

"Clearly, with the absence of knowledge amongst the majority of humans, as you have pointed out, there have been numerous superficial distractions related to the vulnerabilities of the physical body that are strategically created to further keep them away from the truth of their energetic existence.

"As you very well know, the strength of the force of darkness is very much dependent on the quantity of its members. Thus, by doing so they not only energetically increase the number of the dark force, making themselves more powerful, but also further lock humans into their physical bodies, consequently making it even harder for humans to realize their energetic abilities."

"So is it safe to say that when humans such as myself have subconscious dreams instead of conscious travels, this is an immediate sign that our energetics are being attacked, but that we do not have enough consciousness to realize it?" I asked.

"While this does apply to many cases, it also does not apply to a number of cases. In essence, it is very, very difficult to identify which pertains to which, as the human existence has permitted humans to dream and to have nightmares. Therefore, it is very difficult to identify nightmares and dreams as attacks. It is only after experiencing them that a being can use the ability to sense and determine this."

I had spent days and nights researching, ardently seeking answers regarding these attacks, hoping that my understanding of human physiology would be of benefit to finding the truth, only to find myself disappointed every time, for everything that I came across corresponded solely to the physical and completely disregarded the existence of the energetic.

I understood now. All the fearful nights I had experienced, along with my countless travels, were meant to occur in perfect timing so that I would realize the interrelations between the puzzle pieces. From that moment on, I would view everything in its true essence, that of energetics. The human life is but a framed puzzle composed of discrete pieces of paper, glue, and gaps. Though one sees the whole picture, one forgets that the interconnectedness, the points of connection, is what truly matters.

"What about the physical anatomy? Does it matter as much as the energetic?" I inquisitively asked the great being.

"Because consciousness is retained in the physical body as well as in the energetic, the anatomy of the human physical mind [...] is of importance. If certain parts of the physical brain are damaged, then the proper division of consciousness between the physical and the energetic will be disrupted."

Grappling for complete understanding, I flooded the great being further with my longing for knowledge. "My great being, I know that some of these weapons exist in the physical, but how can this be so?"

"These weapons do currently exist in the physical, as there have been realizations of how these negative energies could be developed on physical planet earth. Their primary function rests on their ability to contribute to the negative energies that are perpetrated in the fourth dimension through these attacks."

"What can I share with others about these weapons?" I asked with a hint of desperation. "And what can we do to defend ourselves from them, my great being?"

"As we are all energies and composed of vibrational frequencies, it is very important to stress that though these weapons are not necessarily identified as actual weapons on physical planet earth, they are the darkest and most evil weapons to have ever been created.

"They must work to heighten their vibrational frequencies [quickly], so that electrical currents that target them and attempt to break down their vibrational frequencies are not successful in their efforts.

"As you already know, it is also useful for them to envision themselves being entirely enveloped by the golden light. The golden light is not simply to be perceived as moving from top to bottom, as this succumbs to the human thoughts of gravitational pull and, therefore, exposes one to the vulnerabilities and limits of one's physical body. The golden light is much more powerful and dependable if it is envisioned as entirely enveloping you.

"You must feel the golden light entirely surrounding every inch of your being. With every breath you take, feel it pulling closer and closer into your being and eventually through your mind and through your heart. And let it settle there, where it truly belongs."

A jagged memory of a sensation similar to that of being electrocuted, but with every atom of one's being screaming in agony under the unrelenting pain, surfaced within my mind. It was as painful as in the physical, and the only solace was knowing that this pain could not cause me to bleed or to break, as it would in the fragile physical body. *How could one escape such pain?*

"Every thought should be separated from the aspects of the human physical body and other vulnerabilities, as these are the ones that humans associate most with pain. It is the natural inclination of human beings, whenever they feel any pain, to automatically relate this pain to their physical body. As you very well know, this drastically lowers their vibrational frequency and further encourages thoughts of negativity and [vulnerability]."

"My great being, may I please ask, what is the heart's relationship to the energetics?"

"The rhythmic nature of the heart is what permits the energetic vibrations and higher energetic frequencies of the mind to exist; without the presence of the rhythms of the heart, the existence of these energetic frequencies of the mind would simply not be permitted. And this is precisely why both are pertinent to every being's connection with the power of Togetherness and unity, as the rhythmic nature of the heart is what allows or promotes various beings to join the common rhythm of the power of Togetherness, all existing not only in similar high energetic frequencies, but also in one united sense of rhythm."

Small, rounded blue pebbles like fields of blueberry embraced the planet around me, and as I peered over them, a peaceful, melodic rattling echoed itself faintly through our conversation. There was a clear radiance within these pebbles, a vividness whose essence surpassed their supposed solidity; though they appeared stagnant, even dense, I knew that they held a part within this realm and within this universe, just as the shimmering golden-toned stalks and the lilies did. *What could we do?*

"Again, we need to stress Togetherness and the love because the triumph over these attacks will surely not come from an individual standpoint; it will have to take place in the collective. The thing they are most scared of, as you probably know, is the collective of love and light, and that is the biggest power that can triumph over this, and perhaps the only power that could triumph."

"My great being, I don't understand why the laws of the universe, specifically what humans deem as karma, do not befall them for doing such terrible things."

"The natural universal law unfortunately has been unbalanced and very negatively affected by the presence of very negative and dark frequencies. The natural universal law is, in a sense, encouraged by the presence of high vibrational frequencies and positivity; when such vibrations instead turn to lower vibrations, a critical disharmony is caused, which can be perceived as holes or rifts within the occurrences of the universe. Thus, these occurrences are no longer natural; they are manipulated and caused by the presence of unnatural, lower vibrational frequencies. Therefore, you must understand that the current state of the very process that humans have come to term as karma, though it was originally a process that was naturally created by the universe to promote the unity of beings, is presently of a very, very dark quality."

"What about darkness and light, my great being? Does there always have to exist a balance between them?" This question had long lingered, unanswered, within my mind.

"The unity of universes and the continuity and eternity of universes and galaxies does not [presuppose] a balance of good and evil, or light and dark. We must envelop these forces that we consider dark and so-called evil with love and light, and teach them the power of the Togetherness. It is a great misconception in the many universes, but it is an eternal truth that darkness is not needed."

A strong sense of confusion weighed itself down on me. *If darkness has made its presence known and corrupted most on earth, why haven't the higher beings of the much higher dimensions, such as the great beings of this dimension, intervened?*

"Various attempts of such direct communications have indeed taken place. However, the higher dimensions have been forced to consider other options and methods to continue communicating with beings on earth, as many humans, due to their vulnerabilities, have been entirely taken over by these dark forces, and this puts the higher dimensions and the unification of Togetherness at great risk and in grave danger. You must understand that such communications would allow the dark force to tap into the Togetherness and therefore destroy the only chance left for a positive universe without darkness," the great being explained.

I comprehended immediately what the great being had meant: *because it would corrupt the force of Togetherness.*

"That is very much so," the being agreed. "We do want you to know that such attempts have been made, but the universes and the higher dimensions within them have determined and [agreed] that it would be best to not go about communicating this way, as it would simply put the Togetherness and the higher dimensions at risk and further instill success in the dark forces."

I paused for a brief moment, shaded by the inclination of the group of golden-toned stalks closest to me, pondering over the great being's words and attempting to digest the knowledge that had been shared with me. As they slowly straightened up, symmetrical rays of light fell back upon me, illuminating my view as if encouraging more thought and questioning.

"My great being, are there dimensional divisions within the higher realms, such as in the lower and median realms?" I asked.

"It is very much the case that all beings of the higher nature connect to the force and unity of Togetherness. Though these dimensions are divided, in the sense of their positions within the universe, we are very much united and constantly remain as one; thus, energetically there exist no divisions amongst us."

A light chuckle resonated within my heart, one of joy that basked in the relief of long-awaited realization. How foolish had I been for not noticing these clues—the clues of humanity's true potential, continuously manifested on earth by the ways of nature, and governed by the natural occurrences of the universe. And despite the force of darkness attempting to suppress humans from ever discovering their true energetic essence, these clues have persevered and continue to be manifested by the universe so that we can discover them. *The lilies, the inner circles within a tree's bark, the conjoined segments of bamboo plants...even the transformation of crawling caterpillars into flying butterflies.* Countless times had I chosen to walk around them in aimless circles, carelessly staring outward into the horizon; yet all along, they have patiently remained within the borders of my awareness.

"My great being, it has come to a point, hasn't it?" I finally realized.

"That is very much so. Once humans learn to recognize the innate and universal abilities that are within them, they are given the choice to either join the force of love and light or join the force of darkness, as they possess the key to altering their energetic body in whichever way they desire and also the choice of how they maintain their vibrational frequencies."

"But, my great being, what about those humans who have done such terrible acts? Do they still have the opportunity to join the force of love and light?"

"All humans possess the ability and opportunity to work in the power of love and light and positivity, and to join and become united as one with the power of Togetherness. It is not simply, as one would say, a one-ended road for humans; and although they have been exposed to such negative, evil, and dark origins, it is not necessarily true that they must all remain attached to them."

In the periphery of my vision I caught sight of a twinkling object spinning on its axis. Though its shape and size resembled that of a coin, it appeared to be made up of tiny crystals of ice, each one of them shimmering in a perplexing transparency of different tones of yellow and gold. And as I focused on its rhythmic rotation, I was brought to yet another question: "With so much darkness currently clouding planet earth, I often wonder what the far future has in store for mankind. May I ask for your wisdom of what the energetic probabilities dictate for mankind in the future, my great being?"

"It is very much a future that is led and encouraged by the presence of positivity, love, and light, but as we have explained and as you have discovered, the natural occurrences and natural laws of the universe have been targeted and altered by the dark forces. Therefore, it is up to the beings of today, particularly those living in the third dimension, to identify and at the same time realize that there is this possibility for a future that is immersed with love and light."

A positive future—light will triumph over darkness, and humans of positivity, love, and light will join together energetically to uplift the vibrational frequency of earth through...Togetherness.

"My great being, how can humans be a part of the Togetherness?"

"This can be achieved through the elevation of energetic frequencies, the acceptance of the realities and truths of the many universes and the existence of Togetherness, the realization of their abilities, and the attainment of intense feelings of love and light. Any being is capable of forming a linkage with and becoming part of the Togetherness. We must remind you that you are part of and unified with the force of Togetherness; thus, you will receive assistance in your endeavor."

Within a single heartbeat, I found myself on the peak of a mountain, its sides strewn with vibrant meadows of spring greenery. Spirals of golden light adorned everything in my view, lining the sky boldly and reflecting out into space. But before I could even comprehend what I was seeing, a thought echoed loudly within my mind: *Impossible. I thought I had been somewhere else earlier.*

"Never accept the implications of the word that you state that is called impossible." The message was immediately transferred to the core of my being.

The drizzling of the rain stopped, as I watched a previously closed lily open itself up, its blooming petals revealing its luminous interior in perfect unity. In this moment, I felt an abrupt surge of energy resonate through my being, as though a forgotten melody were being played, its sound making its way within my heart and reminding me of its symphony.

It is through seeds that sprouts blossom and plants are born on earth; but here, it is through positive energies that they evolve.

The message was clear.

"Revival!" I exclaimed.

Energetic revival.

"Come," the great being said. "The natural occurrences of the universe have given you the opportunity to learn more of the Togetherness. Come journey through the many universes, for only experiences alone shall enable one's soul to fully accept and remember what the mind initially could not. The knowledge that has been and will be provided is to be shared with the beings of the universe; even those beings that doubt and turn away from this knowledge would benefit from looking below the surface. It is not difficult for any being existing within this universe, even those living on physical planet earth, to realize his or her energetic abilities; and he or she must be reminded that if he/she just looks closely, proof of it is everywhere. As you well know, the universe and the many other universes and the beings that reside within them are all very much composed of various energies. And if this is not believed by certain beings, particularly those living in the third-dimensional physical realm, then it should be accentuated through the relationships and through the feelings and the bonds that they themselves have noticed them forming in the physical; for these are manifestations of the energetic sense and cannot solely be deemed as occurrences in the physical."

I agreed.

And in Togetherness, we began our journey.

Epilogue

How can you brush off a world of limitless possibilities,
a world of freedom,
a world of love and light,
if you imprison yourself
within a room full of prisoners?

Have you ever sat and wondered who you are?
If you haven't,
sit quietly and ask.
For the heart can only answer
the mind's heartfelt request.

Appendix: The Key Protocols of Multidimensional Travel

* Maintain feelings of love, light, and positivity. Rid yourself of all negativities, daily worries, thoughts of physical constraints, and fear. Keep an open mind.

* Throughout the day, focus on raising your energetic frequency by disciplining your mind with positive thoughts, and take charge of the human mind by shutting out your subconscious "chattering."

* Focus on realizing your own energetic reality along with that of the universe.

* Prior to going to sleep, do the Golden Light exercise to raise your vibrational frequency.

* Before falling asleep, remind yourself of your ability to travel limitlessly and that you are in absolute control of your energetic body.

Glossary of Important Terms

Automated thoughts: Uncontrollable subconscious thoughts and inner "chatters."

Candi: An ancient Javanese temple.

Earthbounds (n.): Beings that were previously humans but have died with an attachment to their prior physical lives, which leaves them stuck in the fourth- dimensional realm.

Energetic diffusion: What occurs when our overall vibrational frequency is lowered as a result of being in the same environment as others who vibrate in a much lower vibrational frequency.

Energetic fuel: A force that is created and elevated through pure thoughts of love, light, and positivity. The production of energetic fuel within the heart is the only possible way to increase one's overall vibrational frequency.

Human energetic realm: Known as the fourth-dimensional realm or the human dream realm. It is within this particular dimensional realm that our energetic bodies exist.

Lower-dimensional beings: Beings who have ceased to possess souls or consciousness. They are in a trancelike state and driven entirely by negativity and hatred.

Soul connection: A connection that persists across time, space, and many lifetimes.

Vibrational frequency: The language of the energetic world. All beings possess unique vibrational frequencies that are composed of distinct vibrations.

Index

About the Author

....

Khartika Goe has evolved from a rigorous academic background into an ardent writer on the unknown and the mysteries of the universe. Ms. Goe has traveled extensively throughout the world and has successfully captured multidimensional beings and planes of existence on film. Many prominent esoteric authors and researchers have used her accounts and photos for their own research.

Her current work focuses on assisting children with physical and developmental disorders in the fostering and cultivation of their energetic abilities. She is in the progress of establishing a specialized school dedicated to teaching these children how to best use their energetic potential. Her Website is *www.thetogetherness.com.*

Katy Kara graduated from the University of California–Los Angeles with a major in economics. She is currently attending law school in California and preparing to undertake her life's goal of assisting abused children and adults around the world through her knowledge of international law. In her spare time, she travels with Goe to film and document their research findings of anomalous occurrences together.

....